LAND
OF
THE POST ROCK

LAND OF THE POST ROCK

Its Origins, History, and People

Grace Muilenburg and Ada Swineford

UNIVERSITY PRESS OF KANSAS

© 1975 by the University Press of Kansas
All rights reserved

Published by the University Press of Kansas (Lawrence, Kansas 66045),
which was organized by the Kansas Board of Regents and is
operated and funded by Emporia State University, Fort Hays State
University, Kansas State University, Pittsburg State University,
the University of Kansas, and Wichita State University

Library of Congress Cataloging Card Number 74-23833
ISBN 0-7006-0194-5 paper
ISBN 0-7006-0129-5 cloth

Printed in the United States of America
10 9 8 7

Acknowledgments

This book's prototype, a mimeographed pamphlet with the same title, was published by the Kansas Geological Survey (1956, first edition; 1958, second edition). Interest aroused by that pamphlet encouraged us to enlarge *Land of the Post Rock* to book form. Soon after the first edition of the pamphlet was released, Jess C. Denious, Jr., late editor of the *Dodge City Daily Globe,* asked permission to present some of the information to his readers. Other area editors likewise responded favorably. The *Topeka Daily Capital* (September 30, 1956) in an editorial prompted by the pamphlet said in part, "Persons who for the first time have visited the county seat towns of Beloit, Russell, Osborne, Lincoln, and Mankato always are impressed with the beautiful courthouses. In driving through the country they also have marveled at the stone fenceposts that surround many fields. The material for building and posts is one of the outstanding features of Kansas mineralogy." C. C. Abercrombie, who established "post rock" as a regional trademark and whose inquiries had inspired the writing of the pamphlet, requested copies to place in guest rooms of his newly completed Post Rock Motel near Lincoln, and when those disappeared he requested more copies. The opening of the Post Rock Museum at La Crosse in 1964 generated additional requests for the pamphlet, some from out-of-state visitors. Others excerpted the pamphlet in publications and brochures, on labels of post-rock souvenirs, and for the sign erected by the Corps of Engineers at the dam site of Wilson Reservoir.

We make no formal dedication because we think people of post-rock country will recognize that the book belongs to them. Regional acceptance of our efforts is our hoped-for reward; we shall be doubly rewarded if persons passing through the area are encouraged to stop to observe and to appreciate the post-rock landscape.

We cannot adequately recognize those who made this book possible by assistance and contributions that in a very real sense entitle them to share authorship. Some persons were sources of multiple information: C. C. Abercrombie, a Barnard banker now deceased, who urged us to prepare a lengthier work and who photographed post-rock subjects for our use; Alan Houghton, *Beloit Daily Call* editor-publisher, who along with C. C.

Abercrombie made inquiries that inspired the mimeographed pamphlet and shared with us his enthusiasm and efforts to encourage post-rock preservation; Margaret Evans Caldwell, Hanston, author of "The Mudge Ranch," who early in the project escorted one of us to post-rock localities in Hodgeman County and arranged interviews with old-timers; Thomas Frusher, now deceased, an early stonemason in the Jetmore vicinity and a faithful correspondent until his death in 1966 at ninety years of age; Ralph Coffeen, Russell artist-historian and ardent promoter of post-rock preservation for more than thirty years, who generously shared information and post-rock exhibits and directed us to fruitful sources of information; Clarence B. Mehl, Jr., Mitchell County farmer and quarry operator who patiently explained the old and new quarrying methods; Harvey Roush, Lincoln County crop reporter for the Soil Conservation Service, who made many scouting trips for us and guided us to photogenic rock outcrops and post-rock structures; Herman Linenberger, St. John Rest Home, Victoria, son of one of the original settlers of Herzog (Victoria) and now believed to be the only living person who worked on the magnificent "Cathedral of the Plains," who provided and checked much information on the building of that edifice; Donna (Mrs. Martin) DuVall, Salina, who made available the unpublished genealogy her mother, Mrs. Allen Webster, prepared on the Webster and Faulhaber families who homesteaded in the Victor vicinity of southern Mitchell and adjacent northwestern Lincoln counties; Harold Dwyer, former newspaperman in Mitchell County and a columnist for the *Hastings* (Nebraska) *Tribune*, who provided much usable copy on the quarrying of stone posts; J. M. (Jake) Herrman, Liebenthal's oldest resident, who recounted for us information on the former use of post rock as explained to him by his father and his grandfather, one of the original settlers of that town; Minnie Dubbs Millbrook, author of *Ness: Western County, Kansas*, who shared transcripts of letters written by an early settler, Mrs. Seymour Mooney, to an eastern newspaper; Harry Grass, president of the Farmers and Merchants State Bank, La Crosse, who passed along stories on post rock's uses as told to him by old-timers of Rush County. We could go on and on, but we must leave space for the text of your book. Under Special Sources in the bibliography, however, we include a roster of those who contributed information by correspondence or personal interview. Illustrations are by the authors, except as noted.

We are indebted to W. W. Hambleton, director of the Kansas Geological Survey, for permission to produce several illustrations and to use material from various Kansas Geological Survey reports; to the Geological Society of America, the University Press of Kansas, the Kansas Highway Commission, and the Kansas State Historical Society for permission to use or adapt for use certain published illustrations; to various members of the State Historical Society and of the area's county and local historical societies for helping us search files for information; to area newspaper editors and publishers—notably Russell T. Townsley of the *Russell Daily News* and Lafe Rees of the *Lincoln Sentinel-Republican*—who uncovered information for us, made contacts for us, or provided old photographs; to area librarians; to officers of county courthouses and city halls; to area postmasters—including especially Lois E. Strnad of Munden, Grace Chegwidden of Bunker Hill, H. C. Cain of Delphos, Mrs. Vernon Felder of Nekoma, and Gene and Virginia Montgomery of Barnard—who helped us find local authorities on subjects in their areas.

Finally we owe immeasurable gratitude to many persons who kept us on course, encouraged us, chided us until we produced. A chief protagonist, Lowell Brandner, redactor for the Kansas Agricultural Experiment Station, instilled confidence and simultaneously almost forced us to write; Floyd W. Smith, director of the Kansas Agricultural Experiment Station, approved such behavior. Frances Swineford, Educational Testing Service, Princeton, New Jersey, and Margaret Oros, Kansas Geological Survey, accompanied us on several data-gathering excursions, read the manuscript, and made many helpful suggestions; Robert Richmond, archivist, Kansas State Historical Society, and Myrl V. Walker, Professor Emeritus of Geology, Fort Hays Kansas State College, reviewed the manuscript, gave valuable guidance, and corrected errors; James Busse, Lawrence, processed many photographs, with a sensitivity that made some usable that might not otherwise have been usable; Helyn Marshall, Manhattan, did the final typing of the manuscript, on a typewriter that surely must be part dictionary; Yvonne Willingham, editor-in-chief of the University Press of Kansas, did the final editing, with a finesse we envy; and John H. Langley, director of the University Press of Kansas, encouraged and goaded us with such inhumane deadlines for copy as Thanksgiving Day and December 26.

Yet another person must be mentioned: Rolla A. Clymer of the *El Dorado Times*. Despite his partiality to the Flint Hills, he has empathy for all Kansas landscapes and for more than a decade has been urging us to expand our booklet *The Kansas Scene*.

Despite our efforts to be accurate, we know readers will find errors, and some undoubtedly would interpret some observations in other ways. Many persons will find gaps in information. We accept those possibilities and the resulting blame and ask tolerance for a fellow pair of human beings.

Contents

Introduction

Ordinarily fence posts would not be considered scenic attractions. But such a distinction, as proclaimed by many travelers, belongs to the creamy-buff, brown-striped stone posts that grace the landscape in north-central Kansas. There, where woodlands are sparse and prairies expansive, man—however inadvertently—has complemented the natural scene by setting out rows of the stately posts, all shaped from the native rock now known as Fencepost limestone, or simply post rock.

The aim of the early settlers who started the practice was not to adorn the landscape but rather to fence their property. To do that, they were obliged to find a substitute for post timber, which was not available on the central Kansas uplands. By chance or by necessity, they began turning back the sod and splitting posts from the rock layer that subsequently was named in honor of its use. By coincidence the finished posts were colorful and when set in place along the fence line blended with the landscape.

Now tourist attractions, the stone posts bespeak of much more, for they were a central factor in the settlement of a segment of the Kansas plains. Though regional in scope, the story of post rock has wide appeal.

Over several decades, beginning about 1880, the use of Fencepost limestone for fence posts had become so well established in north-central Kansas that the settlers' descendants tended to take stone posts for granted. Not until well into the twentieth century, with the coming of better roads and automobiles, was attention refocused on the limestone's regional significance. When travelers began quizzing service station attendants (and others) about the unusual posts, those who had inherited the post-rock land were prompted to become more knowledgeable about their regional trademark, perhaps as much to satisfy their newly aroused curiosity as to accommodate the inquisitive tourists. Becoming conscious of the rock's attractiveness, they also began to see the posts as evidence of their forebears' resourcefulness.

That awakening—plus the realization that wooden or steel posts now are being used predominantly in the area's fence lines, including those relocated because of widened roadways—has stimulated some special efforts to preserve the post-rock heritage. This book is intended to support those efforts. By focusing on the rock itself, we have attempted to meld facts on the rock's occurrence and use. Doing that meant probing into regional geography, history, economy, and structure.

The framework for a century of living on the north-central Kansas plains—beginning near 1870—emerged as we probed into the post-rock history of the Smoky Hills Region. We have been able to record many happenings associated with the use of Fencepost limestone in that drama because one source led to another, which led to yet another. We have described certain associated events because key events relating to post rock's uses could not be told adequately without referring to them. We felt compelled to present post rock in its proper perspective as we observed the geologic occurrence of the area's native limestones and then listened to direct descendants of the area's settlers tell stories their parents and grandparents had told them about their forebears' dependence on post rock to survive on prairie homesteads.

Initially we communicated with geologists familiar with various parts of the Smoky Hills Region, and we consulted pertinent geological literature. That provided much ancient history plus clues on resource exploitation, which directed us to associated human history. Simultaneously we received direction from letters received at the Kansas Geological Survey from area residents, who at first were the inquirers: What can you tell us about the occurrence of this rock? What makes it suitable for use as posts and building material? The procedure reversed as we became the inquirers: When did the pioneers begin to quarry the rock? Who started the stone-post routine? Why? Our correspondents began suggesting other area residents to contact. That we did by letter and soon by visits into the area, where we queried anyone who happened to be "on the spot"—in a farm home, in a small-town store, in a newspaper office, in a rest home, at a service station, in a courthouse or city hall, in a cafe, in a chamber of commerce office, in a public library, in a post office, on the street—near where we wanted to photograph and acquire information on a row of stone posts, a stone building, or other stonework that appealed to us.

The mimeographed pamphlet *Land of the Post Rock* was based mainly on material collected from Kansas Geological Survey publications, correspondence with a few area residents, and a few excursions to observe post-rock outcrops and the stone-post–decorated landscape and to talk with a few residents. The idea of a book took form between 1956 and 1970. As our file of information grew, we attempted to separate fact from folklore but did not completely succeed—which may be just as well, for who can conceive of any heritage that does not include a bit of folklore?

We spent many hours, at different times, at the Kansas State Historical Society, searching through files of newspapers published through the years in the region's counties and towns and examining bibliographies, reference works, and indexes to Kansas and regional publications. We also consulted the Special Collections of Farrell Library, Kansas State University, and the Kansas Collections, now housed in the Spencer Research Library, University of Kansas. References led to other references. Useful material was found in area libraries and historical museums. More than once a letter sent to an individual whose name came to us through chance or as a "possible suggestion" resulted in unexpected returns.

Major sources of published information are listed in the bibliography, rather than distract the reader with numerous footnotes or parenthetical documentation in the text. Most important of all, our contributors—secondary authors—are listed in the Special Sources section of the bibliography.

Let us, then, enter the Land of the Post Rock.

LAND
OF
THE POST ROCK

Landscapes,
except where marred by human maltreatment,
show perfect harmony in all their parts,
and I submit Kansas,
without lofty mountains or awe-inspiring canyons,
as a thing of beauty composed of placid forms.
It was created less violently than some other parts of the world
but by forces just as relentless and just as exciting.

— John Mark Jewett,
The Geologic Making of Kansas

1
A GEOGRAPHIC SETTING

Driving west in Kansas on Interstate 70, we reach the Land of the Post Rock about midway across the state, or approximately thirty miles west of Salina. That thirty-mile stretch after Salina we can consider our approach to post-rock country. The approach is in the easternmost part of a plains border-land known as the Smoky Hills, and as we move through it we can scarcely ignore the rugged topography accented by brown sandstones protruding from the hilltops. Dakota Country some call it, because those prominent sandstones belong to a geologic unit of rocks called Dakota formation.

A few minutes after we cross the Saline County line into southeastern Lincoln County, we begin to penetrate post-rock country. Also a part of the Smoky Hills, that area exhibits broader uplands than are seen in Dakota Country, and chalky limestones are the eye-catching exposures in the hill-sides. To establish our approximate entrance, when the sign "Enter Lincoln County" is visible, we suggest a leftward glance, toward the southwest, for a view of this landmark: a barn constructed in 1880 of brown sandstone (from the surrounding sandstone hills) and trimmed with creamy-buff Fencepost limestone, or post rock (from the upland slopes to the west).[1] Soon after the barn is out of sight, we come to the Carneiro-Beverly exit, our cue to begin looking for a few stone posts in fence lines on the countryside.

After traveling seven miles in Lincoln County, we enter Ellsworth County. Near the roadside park, just past the U.S. 156 exit, we can distinguish ahead an east-facing escarpment (rather steep slope) trending irregularly in a northeast-southwest direction. That escarpment, built up of rock layers of the Greenhorn limestone, with the famous Fencepost bed at the top, is the true eastern boundary of the Land of the Post Rock.

At the roadside park (rest area), a good place to stop for orientation, we

Dakota Country and transition into Land of the Post Rock. In vicinity of Kanopolis Reservoir, Ellsworth County (upper photograph), and in Hell Creek area southeast of Wilson Reservoir, Lincon County.

Sandstone barn with post-rock trim, on the Andrew Yordy homestead in the southeastern corner of Lincoln County, as it appeared in the late 1890s or early 1900s. *Photograph courtesy of Helen Craig Dingler.*

can observe posts placed there by the Kansas Highway Commission in 1964. And the inscription on the historical marker placed there by the Kansas State Historical Society alludes to the past use of stone posts in north-central Kansas.

The Land of the Post Rock stretches about two hundred miles from the Nebraska border near Mahaska, northwestern Washington County, almost directly southwest to a few miles north of Dodge City, Ford County. East to west, the boundaries of the area so zigzag that its width ranges from less than ten to more than forty miles (to approximately sixty miles if measured along Interstate 70). Roughly it covers five thousand square miles, or more than three million acres.

Included almost entirely in the Smoky Hills physiographic province, post-rock country proper, where Fencepost limestone crops out, comprises much of or parts of these counties: Republic, Jewell, Osborne, Mitchell, Cloud, Ottawa, Lincoln, Russell, Ellis, Ness, Rush, Barton, Ellsworth, Pawnee, and Hodgeman. Fencepost limestone's use for posts, however, extends beyond the rock's outcrop area. Plausible extensions are into extreme northwestern Edwards, Rice, and Saline counties and extreme southeastern Rooks, Smith, and Trego counties.

Immediately to the west of post-rock country, a rock unit known as Fort Hays chalk, which in texture resembles Fencepost limestone, has been quarried to some extent for posts. Thus, the westward extension of the post-rock area is not well defined. Continuing west on Interstate 70, however, when

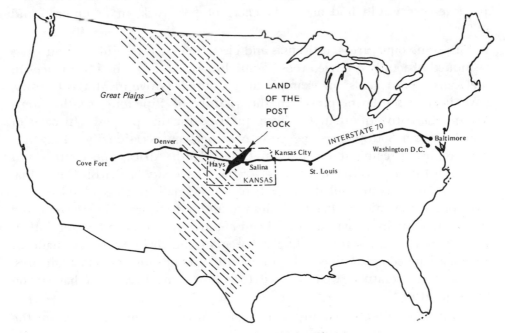

A unique area in mid-America.

A Geographic Setting

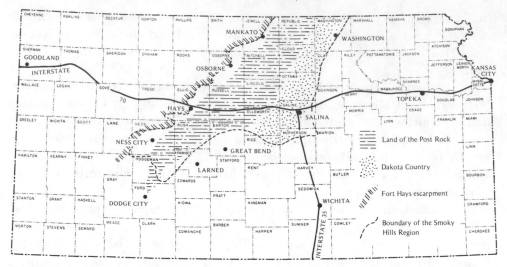

Land of the Post Rock, showing position in Kansas and in the Smoky Hills Region. Dakota Country (eastern approach) and Fort Hays escarpment (approximate western boundary of the Smoky Hills) also indicated.

we reach Hays we will be at the approximate western limit of the area where post rock has been used. Should we leave the superhighway, by way of any exit between the Ellsworth rest area and Hays, for jaunts on lesser highways and section-line roads—north and northeastward toward Lincoln, Beloit, and Belleville or south or southwestward toward Hoisington, La Crosse, and Jetmore—we would find many evidences of post rock, in its outcrop and in use.

In some other areas, in Kansas and elsewhere, it is possible to find posts fashioned of locally quarried rock. Some limestone ledges in a few places in the Kansas Flint Hills, for example near Council Grove in Morris County, have been a source of posts. In the vicinity of Pipestone, southwestern Minnesota, native quartzite has been quarried for that purpose. In parts of the American Southwest stone posts may be seen occasionally. Travelers have reported seeing stone posts in India, South Africa, and central and northern Europe. William Mathews, of the University of British Columbia, noted stone posts around sheep pastures on a small, treeless island off the coast of Norway midway between Ålesund and Sognefjord. He also reported similar fencing in a small area of sheep country in southern Switzerland in a valley north of Locarno. (The Norwegian and Swiss posts are made of gneiss, a banded crystalline rock formed under conditions of very high pressure and temperature, well adapted to post-making because it has strong linear jointing.)

But it is unlikely that any other region anywhere can claim, as can the Kansas post-rock area, a single bed of rock that has been used so extensively

Topography typical of post-rock country. Scenes in the vicinity of Wilson Reservoir, Russell County.

for fence posts that the posts have become an identifying feature of the landscape.

Why should such a distinction belong alone to about three million acres in north-central Kansas? The reasons become evident as our story unfolds and we place in perspective all factors involved: the area itself (its geology, physical characteristics, natural resources) and the people who settled there (regional history and development).

The rest areas along Interstate 70 north of Ellsworth are at a point of transition in topographic expression.[2] Rugged Dakota Country, developed on relatively hard sandstones and softer clays and shales, meets the Land of the Post Rock, where the landscape drapes flowingly over hillslope benches of Greenhorn limestones and semi-protected shales. For reasons to be discussed later, some geologists and geographers include both areas in the Smoky Hills Region; others use "Blue Hills" to refer to the approximate area where stone posts are quarried and to a strip of plains borderland on the west.

Whether we consider the Land of the Post Rock as a part of the Blue Hills subdivision or simply the central part of the Smoky Hills (with the Blue Hills included), we can be sure that the surface we now see owes its form to the differential wear and weathering of the underlying rock strata (together called the bedrock). Consequently, during the last one or two million years water, wind, and other natural agents have been eroding and carving the bedrock and breaking rock masses into bits and particles that become a part of the soil cover.[3]

We leave until later what the makeup and structure of the bedrock have to do with the weathering process and shaping of the landscapes. For now we must be content with this observation: Because they tilt slightly downward from east to west and because they are "broken off" where they protrude at the surface, the region's (indeed the state's) rock strata are arranged like a slumped deck of cards—the oldest layers, those laid down first in ages past, are exposed toward the east; the youngest, toward the west. Yet the surface itself is an incline with its highest altitudes toward the west. Elevations in post-rock country, for example, range from 1,400 feet above sea

Block diagram of Kansas, showing surface incline to the west, generalized contours indicating above-sea-level elevations at 500-foot intervals, and post-rock country (hachured area).

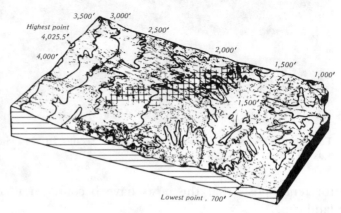

level along the Republican River in the northeastern part, in Republic and Cloud counties, to about 2,500 feet on some divides in the southwestern part, in Ness and Hodgeman counties. We can stand on the prairie upland, however, and not be aware of this incline, for the rate of rise is only about ten to twelve feet a mile (east to west).

Locally, on the other hand, considerable ruggedness may be evident in short distances. For example, along the Smoky Hill River, the upland surface commonly is 150 feet above stream level, and the Saline in places cuts 300 feet below the upland. In such places—that is, along stream banks, in road cuts, and in ravines where vegetative cover is lacking—we can observe, in cross section, rock strata that form the area's uplands: principally alternating layers of limestones and limy shales of the Greenhorn limestone sequence. In places where the upper part of the Greenhorn unit is well exposed—for example, in the Wilson Reservoir area—the topmost bed, the relatively thin Fencepost limestone layer, may appear as a scalloped edging along the crests of the ravines or roadside exposures.

Once perceived as a vast, treeless grassland incapable of habitation, the Land of the Post Rock since the 1860s has been transformed from free range

Along US 183 north of La Crosse, Rush County.

to productive farmland. The use of Fencepost limestone for posts, and for building material as well, had much to do with the transformation. Though now displaying a modern transportation network and other exhibits of man's presence, the region remains rural and agricultural.[4]

The only town having a population approaching 15,000 is Hays at the western edge, where Fort Hays Kansas State College and the largest branch of the Kansas Agricultural Experiment Station are located. The largest towns in the area proper are Russell (about 10,000) and Concordia (about 7,000); Belleville, Beloit, and Hoisington are the only other towns of more than 3,000 residents. "Gateway" towns south of the area include Dodge City, Larned, and Great Bend; Salina, about thirty miles east of the eastern edge, is a major trading center for the area.

Chief farm crops of the post-rock region are wheat, sorghums, and hay (alfalfa and wild), and in some parts also corn and barley; and livestock raising remains important throughout the area. That agricultural combination has been found adaptable to the climatic extremes of the plains border, where annual precipitation averages twenty-six inches, although more nearly twenty-two inches in the western part; average annual temperature is between fifty-three and fifty-six degrees Fahrenheit; the growing season, from the last spring frost to the first fall freeze, generally from late April to about mid October, is about 170 days; and wind velocity is moderately high and evaporation rapid in the warm season.

As we see when driving along some of the byways, between alignments of stone posts, the post-rock motif complements the rural, agricultural setting. Within the geometric design of stone posts about the pastures and fields are many buildings, bridges, and other structures made of the native

Fencepost limestone. Yet enough open spaces remain to remind us of what the virgin prairie must have been before it was crisscrossed by stone posts and plowed fields. Imagine, for example, a sea of buffalo grass, here and there mixed with taller grasses and seasonally colorful with sunflowers, plains indigo, prairie primroses, and other wild flowers, all bowing repeatedly in the wind and rippling away to the horizon. With not a tree in sight. Imagine prairie chickens, jackrabbits, and other small range creatures moving about in search of food. Listen for a chorus tune-up: the crisp note of the meadowlark, chattering in prairie-dog town accompanied by the raspings of the grasshoppers or the thundering crescendo of an approaching buffalo herd; then at dusk the piercing howl of the coyote.

After such reflection, it is well to move on, to descend into a ravine where native woodland hugs the river banks: cottonwood likely, willow probably, ash perhaps, hackberry maybe, and undoubtedly others. Glancing up, if we have been lucky in our choice of ravines, we may notice near the top a uniform, relatively narrow, buff ledge with the brown streak characteristic of the Fencepost limestone bed. We may also see a chalky white patch and maybe a split-out post nearby, which would indicate an old quarry. A walk to the site should give us a chance to look at the rock in its natural setting and to speculate on how the post was split out.

A few more steps to the shelf above the Fencepost bed will give us a vantage point to a broad view of the now productive but once pristine landscape. Moreover, noting that we are no more than a foot or perhaps eighteen inches above the post-rock layer, we see that it was a readily accessible resource to pioneers. Looking across the ravine to surrounding uplands, we can observe a wide, flat area underlain by the post-rock ledge. Simultaneously, we may well marvel at the miles and miles of stone fence posts. With few other building and fencing materials at hand, perhaps it was inevitable that the early inhabitants would discover the potential of this accessible ledge and open up quarries all along the hillslopes.

Stone post above outcrop of Fencepost limestone, along road north of Wilson Reservoir, Russell County.

2
LAND OF THE
POST ROCK ABORNING

When Dr. Brewster H. Higley in 1873 sat on the bank of Beaver Creek in Smith County and wrote the words to "Home on the Range," he recorded a waning Kansas scene.[1] West beyond the sixth principal meridian, or approximately the western two-thirds of Kansas, a few buffalo still roamed and the deer and the antelope continued to play. Wild turkeys, prairie chickens, and a variety of small furry animals at home in grassland habitats abounded. The skies were unclouded most of the time, and except for an interlacing of wooded ravines the treeless landscape stretched on and on. The buffalo and grama grasses and associated vegetation, having developed complicated root systems, which tended to keep the soil in place, were adapted to the relatively dry, windswept plains. A short time before, and for several centuries before that, the American Plains Indians had been happy in such a hunting ground.

But by the 1870s the railroads had begun to cross the plains, and Americans of European descent were staking claims in the short-grass country. Soon the buffalo would be exterminated and the Indians removed from their hunting paradise. In a plains-border section in north-central Kansas, not far east of Higley's "Home on the Range" cabin, the Land of the Post Rock was aborning. There, within a decade, dugouts and sod houses would begin to give way to substantial dwellings, constructed of stone from a rock layer that for hundreds of centuries had lain dormant under the prairie sod. Eventually almost every hillside would give up some of its rock for building

11

"Pristine" landscapes in the Land of the Post Rock. South of Lovewell Reservoir, eastern Jewell County, in the area's northernmost part (upper photograph) and near Bazine, eastern Ness County, in the southern part.

material to be used on the farms and in the towns springing up along the incoming railroad lines. Simultaneously the countryside would become a patchwork of fields and pastures, made increasingly conspicuous by stone posts supporting fence lines about them.

Stone in use thus became a sign of permanence and progress in the area. Consequently, any story of the area's settlement and development without adequate reference to the resource that provided so much building and fencing material would be incomplete. Likewise, to report on that resource and its uses without relating it to the area's history and development would be to tell a story out of context. Historical background thus becomes a prerequisite to an understanding of the regional development.

GOING WEST—TO KANSAS

The year after Kansas became a state, Congress passed the Homestead Act of 1862. That act, extending the privileges of the Preemption Act of 1841, was to have a major impact on the state's settlement. Under either the pre-emption or the homestead law, any eligible person—a United States citizen or one who had declared intent to become a citizen and who also was a head of a household, a widow, or a single man (or woman) over twenty-one years old—could acquire up to a quarter section (160 acres) of government land for certain minimal considerations. Under the Preemption Act, an individual could settle on the land and make some improvement before filing; then, within a certain period—a year or thirty-three months, depending on whether or not the land had been "offered" at public sale—he gained title by paying the minimum government price, $1.25 an acre, or if within ten miles of a land-grant railroad, $2.50. Under the Homestead Act, the settler staked his claim, paid the legal fee ($10.00) and half the commissions ($4.00) for 160 acres "held at $1.25 an acre" or 80 acres "held at $2.50 an acre." He then was allowed six months to take possession by occupation and improvement. After living on the land and cultivating it continuously for five years, he appeared within the next two years at the district land office with two witnesses to offer "proof" of settlement, paid the remaining commissions ($4.00), and received his certificate of title and patent. A settler who didn't want to wait five years to obtain title, and who was fortunate enough to have $200 cash, could buy his claim after six months of residence, provided he had made suitable improvements on one or more acres.[2]

During the Territorial Period (1854–1861), Free-State and proslavery advocates were feuding in eastern Kansas, and many "squatters" had land speculation in mind even as they proclaimed their aim to be to determine whether Kansas would enter the Union as a Slave or Free State. To complicate matters, many Indian tribes from eastern states had been removed to reservations in eastern Kansas, where they supposedly were to live undisturbed forevermore. Those promises and agreements, of course, were made when the white man was still under the delusion that Kansas was mostly "barren, fit only for habitation by roving Indians." By the time Kansas was made a territory, the government had begun "negotiating" with the Indians, the aim presumably to return reservation land to the public domain. But in setting up trust lands through a treaty process, the government established a policy that permitted Indian land to be transferred (through public auction or other means) directly to private ownership without ever becoming a part of the public domain. In any case, in the early territorial period, trust lands were still unsurveyed, titled land was not easily distinguishable from untitled, and much of eastern Kansas could not be homesteaded legally.[3]

Soon after Kansas joined the Union as a Free State, the Civil War broke out, further complicating the westward movement. Even so, the message "Go west, young man" kept showing up in the agricultural sections of Horace Greeley's New York Tribune.[4] So by the time the Civil War was over, many would-be homesteaders had their sights on free land in Kansas. A number were war veterans ready to take advantage of new homestead laws granting veteran preferences, including reductions in residence requirements and the right to 160 acres (limited to 80 acres for other homesteaders) of double-premium, reserved public lands along the railroad lines. (By 1872 a law had been passed that permitted a soldier to deduct his years of military service, up to four years, from the five-year residence requirement. He also could file on land through a land agent.)

Most of the state's free land was in central and western Kansas. Settling there, however, had its drawbacks. Except for a few forts along the trails across Kansas and the route staked out for the Kansas Pacific (Union Pacific) Railroad, the Kansas frontier in the 1860s was unprotected. The plains and nomadic Indians, including the Pawnees, Cheyennes, Kiowas, Comanches, Arapahoes, and Sioux, were yet unwilling to relinquish their "nation," which extended as far east as western Republic County. They made depredations on settlements during that period.[5] Moreover, geographies long had shown the treeless Kansas prairie as a part of the Great American Desert, with the result that many potential settlers were skeptical about its future.[6]

Paradoxically, by the late 1860s cattlemen, who had discovered that stock could subsist on the nutritious prairie grasses without supplemental feed, coveted the grassland as free range. About the same time, the Kansas Pacific Railroad reached the area and a succession of cow towns, or shipping points for droves of cattle from Texas, appeared. Those droves trampling over property threatened the chances of permanent settlements, especially in view of the few resources at hand for fencing material.[7]

By 1870 the migration fever had intensified until the desire to own a piece of land offset the imagined and real drawbacks to settlement in the plains borderland. The pioneers began coming, predominantly from nearby states to the east and northeast; according to the *Fifth Biennial Report*, State Board of Agriculture, largest representations were from Illinois, Indiana, Missouri, Pennsylvania, New York, and Kentucky. A few Negroes came from the Deep South. Besides the lure of the land itself, there were other inducements. Some saw the chance to start anew, for no one could be forced to sell property obtained by homesteading to pay existing debts. Capitalists came to invest and speculators to "get rich quick." Some, plainly adventurous, responded to the excitement of the frontier.

During the season of their arrival, many decided they had found a land of opportunity. Their letters to friends and relatives brought in more settlers, as did promotional literature distributed by special-interest groups and government agencies. Much of the literature was overly optimistic.

Railroads were especially influential. If the railroads had not penetrated central Kansas in the late 1860s and the 1870s, the Indians probably could have held their hunting grounds somewhat longer. Both the timing and the circumstances of their coming made the railroads a powerful force in settling and developing the area. Contrary to the usual situation, the railroads preceded much of the settlement. Their push across the sparsely inhabited plains was spurred by generous land grants and low-cost loans from the federal and state governments, as well as bond subsidies by counties and municipalities. The grants provided for a right-of-way that included two hundred feet of ground on either side of the track for stations and buildings and in addition alternate sections of land up to twenty miles on either side.

As they charted their routes and laid their rails, the railway companies ensured their own future by encouraging not only agriculturalists but also industrialists and town builders to buy tracts along their lines. This they did in various ways. Their promotional brochures advertising land opportunities also gave glowing accounts of the region's natural resources and other assets. Realizing that a viable agricultural economy was necessary if they were to prosper from shipping goods and products and transporting passengers, they sold land at low prices and established low freight rates on certain types of products. For example, the Kansas Pacific Railway, according to the *Fourth Annual Report* of the State Board of Agriculture, in 1875 priced its land at a minimum of $2.00 and a maximum of $6.00 an acre. The Atchison, Topeka & Santa Fe Railway Company listed a price range of $1.75 to $12.00 an acre. Those prices, however, did not remain stationary from year to year.

Few who homesteaded or bought land in central Kansas after 1870 saw an Indian raid. Nonetheless, all who lingered for at least a few seasons faced other disasters: droughts, blizzards, tornadoes, prairie fires, grasshopper invasions (a major one in 1874). Some, finding not what they expected but rather a hostile, lonely, or unfamiliar environment, returned to their former surroundings. Others persisted, if not from sheer determination, then because they had no alternative. But for whatever reasons, so many came and stayed that before the end of the 1870 decade most counties in central Kansas had been organized.

A NEW HOMELAND FOR EUROPEANS

Many of those who had no alternative but to stay were European immigrants. Their greatest influx into post-rock country was during its formative period, the 1870s. In addition, many settlers who migrated from the eastern states were American-born sons and daughters or grandsons and granddaughters of the foreign born. So of those who listed other states as their birth places, undoubtedly more than half were keenly aware of their European origins.[8]

In the mid–nineteenth century, living conditions in many European countries were deplorable not only because of natural misfortunes that upset the population-food ratio but also because of economic changes and societal injustices. In England, for example, the Industrial Revolution although improving the lot of some left little hope for many others, and the revolution in agriculture literally robbed farm laborers of their livelihood. Many of the poor left England, Scotland, Ireland, and Wales for the uncultivated land of the American West. Ironically, at the same time, English capitalists, seeing investment opportunities on the American frontier, also came or entrusted their moneybags to their representatives.

In the Scandinavian countries drought and famine underscored thoughts of migration. In the heart of the European continent, several decades of economic distress growing out of the Napoleonic wars were being extended by events that culminated in the Crimean War. To add to the general unrest, the masses, especially in Germany, were picking up the delayed signals of the French Revolution, resulting in a trend toward liberal ideas, in conflict with existing political repression and religious discrimination. Furthermore, Germans who, because of their beliefs, had fled to more understanding European countries were then oppressed by new rulers in those countries. Germans who had gone to southern Russia were a notable example. Understandably, such peoples were in a mood to listen to any talk of emigration to a "land of opportunity."

Coincidentally, information on the American frontier was crossing the Atlantic in letters, travel accounts, guidebooks, promotional pamphlets, magazines, and newspapers. Various American groups, including churches and emigrant aid societies, were prepared to give financial aid or to assist in bringing groups to this country and perhaps also to help locate them somewhere on the public domain. Two of the active societies were the American Emigrant Company and the Columbian Emigrant Company; both had extensive operations in Europe. The Kansas legislature of 1864 established an immigration bureau, albeit without appropriating funds for the bureau's work. Yet the bureau prepared an "immigration letter paper" and circulars, which were distributed widely and included an appeal for developing financial resources to encourage immigrants to come to the state.

Of all the promoters of immigration, the railroad companies had taken the lead in the 1870s. They printed pamphlets in foreign languages to distribute in major European cities and at debarking points in the United States. Special agents went abroad and special agents greeted new arrivals to America. Representatives overseas sought desirable settlers and offered them low-cost land and free or low-cost transportation to the available land. As presented in a table on page 666 of the *Fourth Annual Report* of the State Board of Agriculture, records of the Atchison, Topeka & Santa Fe Railroad Company, particularly active in such endeavor, showed that except for 752 purchasers from Illinois, more of that railway's land purchasers (533) in

LAND OF THE POST ROCK

Kansas between 1873 and 1875 had come directly from foreign countries than from any state in the United States.

Simultaneously, interest in agricultural education had leaped forward, and indeed had much to do with the passing of the Homestead Act and the companion piece of legislation, the Morrill Act, which established state agricultural colleges and set aside special sections of the public domain for schools and other public services or to sell for various educational revenues. Kansas State Agricultural College (now Kansas State University), Manhattan, established in 1863, was one of the first land-grant colleges.

The interaction of events and their timing were such that the United States—and Kansas—in the 1870 decade received the greatest influx of foreigners since the American Revolution. In the same decade post-rock country, a part of the West just opening up and becoming accessible, became the new homeland for many of those immigrants.

Scandinavians were among the first foreign arrivals. In 1868 three Scandinavian land companies organized in Illinois to bring colonies to

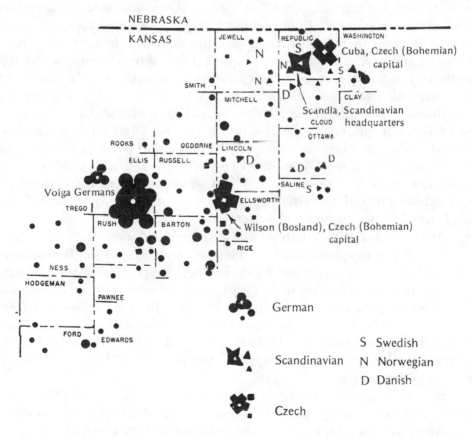

Principal foreign settlements (not including English, Irish, Scotch, Welsh) in post-rock country. Relative importance indicated by size of symbol. *Data from J. Neale Carman's Foreign Language Units of Kansas.*

Kansas. Two, the First Swedish Agricultural Company of McPherson and the Galesburg Company, were responsible for major Swedish settlements in the Smoky Hill Valley in McPherson and Saline counties, adjacent to post-rock country on the east and south. The Scandinavian Agricultural Society of Chicago, of most concern here, secured twelve sections of land in the Republican Valley in southwestern Republic County, and soon afterward the first group of Scandinavian settlers arrived there, traveling on the Kansas Pacific as far as Junction City and then on foot. At a site they called New Scandinavia (Scandia after 1876) they erected their headquarters and fort: Colony House, a three-room frame structure. To make the walls of their "fort" less penetrable in case of an Indian attack, they placed "between a double row of two by six uprights . . . six inches of rock" (*History of Republic County, 1864–1964,* compiled by Anona Shaw Blackburn and Myrtle Strom Cardwell, pages 11, 248–281).

Few Scandinavians who came to the colony and located in the immediate vicinity of Scandia had encounters with Indians. Several Swedes, however, ventured north along the Republican River and west along White Rock Creek into Jewell County, where angry aborigines appeared often enough to abort their first attempts to claim land in this northern part of post-rock country. Among the unhappy Indians were survivors of the Republican band of the Pawnees, a farming and hunting people who until well into the 1800s occupied earth-lodge villages on the bluffs along the Republican River in southern Nebraska and northern Kansas. Trappers, who followed explorers up the Missouri River and across Pawnee territory, by mid century had very nearly wiped out the several bands of this tribe by distributing among them bundles of clothing infected with smallpox and other contagious disease organisms. When we add such deeds to white man's encroachment on red man's hunting ranges, we can begin to understand why the Plains Indians resented the settlers.

After 1870 Indian troubles subsided, and substantially built stone dwellings began to appear on the divides west and south of the Republican, where Swedes, Norwegians, and Danes had become established. From stone quarries and lime kilns on their farms, these Scandinavian craftsmen obtained building blocks and mortar not only for their dwellings and farm structures but also for their schools and churches and business houses. E. D. Haney, in "Experiences of a Homesteader in Kansas," called the Scandinavians "thrifty, in the beginning poor, but now [1916] some of the most wealthy in the county."

In 1869 two major Danish settlements, still viable in the 1970s, took root in post-rock country. One, at Jamestown, Cloud County, was composed of Baptists (few Danes then, as now, were Baptists). The other, at Denmark, Lincoln County, was composed of Lutherans; it remains the most characteristic of the Danish settlements in Kansas. Stone churches built at both locations during the first decade of settlement are the most visible links of both groups with their past.

The Excelsior Colony, consisting mainly of Scotsmen and Englishmen who knew more about stone cutting and other crafts and trades than farming, entered post-rock country by way of Republic County in 1869. Among them were merchants, clerks, professional men, mechanics, and craftsmen (including stonecutters, stonemasons, and stonesetters). After landing in America, they had worked for a while at their various trades in New York City, where they organized the Excelsior Colony. They came to Kansas in three divisions. The first, sixty persons, in May of 1869 attempted to settle along White Rock Creek in Jewell County. The Indians, however, were no friendlier to them than to the Swedes, so even though they had built a fort near Holmwood, several miles east of the present Burr Oak, they had dispersed to safer parts of post-rock country before the year was out.[9]

The second division of the Excelsior Colony, consisting of seven stonemasons, one painter, and the wife and child of one of these men, came in late 1869 and filed claims along Rose Creek north of Belleville. Working together, they quarried rock and built for each a simple stone house. Presumably they learned to farm well enough to prove up their claims. And they soon found plenty of work building stone houses for others.

The third division late in 1869 sent two men to "spy out the land." The men decided on the Scotch Plains southwest of Belleville, mainly because of the "native, soft limestone rock" there. In 1870 the division—seven married couples (several with children) and twelve bachelors—arrived to file their claims. After building sod houses, they built (on the corner where four homesteads joined) a stone house for protection against Indians. Later, they called it "Fort Nonsense," because they never needed it. As they learned to farm and their living improved, they built elegant stone houses and barns, some of which are still extant in southern Republic and northern Cloud counties.

The state's two Bohemian capitals, Cuba and Wilson, are in the Land of the Post Rock. As early as 1866, a few Bohemians (Czechs) began coming to Republic County, "to provide a better life for themselves and their children" than was possible in the military-dominated old country (Blackburn and Cardwell, *History of Republic County,* pages 106–140 and 177–228 especially). They settled mostly in the eastern part of the county and in adjacent Washington County. The towns they helped to found included New Taber (no longer extant), Narka, Agenda, and Cuba. The Czechs were hardworking people, and among them, as among the British and Scandinavians, were those who had been stonemasons in their native homeland. By the time they had patented their homesteads many had stone houses and stables, dug wells, and orchards. Within a decade they constituted half the population of Cuba, which became so much a center for their activities the town became known as the Czech capital of Kansas.

Another Bohemian community in post-rock country, however, arose to claim that title: Bosland, now Wilson, above the Smoky Hill River in northwestern Ellsworth County. It emerged as a Bohemian center soon after Francis Swehla arrived there in 1874 and began working diligently to find homesteads for his countrymen. The "Bohemians around Wilson" still hold annual Czech festivals and they continue to advertise Wilson as the Kansas Czech capital.

Swehla, though born in Bohemia, spent his boyhood in Iowa and while still in his teens fought in the Civil War. So when, in 1868, he and his bride headed a caravan of prairie schooners west, Swehla came both as a foreigner and as a veteran. The caravan stopped in Saline County, Nebraska, and after he helped to build up a Bohemian settlement there, he decided to move south to find another location for a Bohemian colony where land was abundant and free or cheap. At Wilson he found a friend who was a fellow veteran of the Civil War, so he decided to build his "empire" there. As he noted in "Bohemians in Central Kansas," page 474, he began writing letters and reports on everything in the area's favor: ". . . temperate climate; good soil; free land from Uncle Sam, or cheap relinquishments of improvements by previous settlers; railroad land at from $2.50 to $5.00 per acre; good and plenty of water from never-failing springs and wells from thirty to sixty feet; plenty of building stone of fine quality, and an accessible railroad station. A paradise for poultry, cattle, horses, sheep, hogs."

His reports, published in Bohemian-American newspapers, drew the attention of "farm clubs" in eastern cities and precipitated many letters of inquiry that eventually brought numerous Bohemians to the area. Bohemian settlements fanned outward from Wilson: up the Saline into Russell and southeastern Osborne counties and into Lincoln County as far as Sylvan Grove; south and east into Ellsworth and southwest into Barton. Swehla's persistence in bringing Bohemians to the area and in finding land for them seems especially remarkable when one considers that the great grasshopper disaster of 1874, drought, and other discouragements had triggered an eastward migration. Somehow the Bohemians persisted, even though most were more skilled in trades than in farming. Their abilities, however, included skills essential to provide implements of sorts to turn the tough sod into cultivated land and to establish homes where building materials were scarce. Especially needed were blacksmiths, masons, carpenters, and well diggers. Swehla's many references to stone quarries and to the use of stone for permanent buildings attest to the importance of rock as a resource in this community.

Victoria, in the western part of post-rock country, emerged in the 1870s as the unlikely setting for a study in contrasts in foreign settlements. There one venture, backed by wealth, failed; another, with little or no wealth, succeeded. First, the one that failed.

George Grant, wealthy Scottish merchant, became America's number one landholder in 1873 when he bought from the Kansas Pacific Railroad about seventy thousand acres of land in southeastern Ellis County.[10] He had a grandiose scheme: to create in the heart of America a cultural community peopled by English aristocrats and gentlemen agriculturalists. First he laid out a town and named it Victoria in honor of his queen. He induced the Kansas Pacific to build a two-story stone structure, Victoria Manor, to be used as a depot and as temporary living quarters for the scions of wealthy Englishmen he was to bring over to get his venture started. He then persuaded his wealthy English acquaintances that Victoria was a place in which their sons could add to the family fortunes while living in a wholesome, rural environment; besides he would sell them land for an average of 2£ sterling, or about $11.00 an acre (Grant had paid approximately eighty-eight cents)—much cheaper than buying more land in England. Further, he (Grant) would plow the land for them—with cultivators imported from England. Additionally, the sons would be his guests at the depot-hotel until houses could be built. He convinced the young men that life could be exciting and hunting on the prairies glamorous. The first group of young aristocrats, some with wives, set sail April 1, 1873.

After arriving they seemed to be content to live off their remittances, having little desire to farm or raise blooded livestock—interests that Grant had hoped to develop once he got them there. Rather, they preferred to don their hunting jackets, mount their imported steeds, and with their imported hounds alongside, gallop across the prairies in chase of coyotes and jackrabbits. For another diversion, they presumably found the energy to dam Big Creek, using native limestone, to form a lake between Victoria and Hays; they then imported a steamboat, which they sailed on the lake until a flood broke the dam. (William Baier of Victoria, in a recent letter, pointed out that although he would not say that the steamboat-dam story is not true, he finds it hard to believe, considering the terrain of Big Creek.)

Gradually they grew weary of prairie life. Thus, despite the available wealth and ambitions of Grant, who brought the first Black Angus cattle to America and who seriously worked to set up a stock-breeding farm south of Victoria, the colony failed. After Grant's death in 1878, most of the fun-loving noblemen returned to the British Isles, and another colony of foreigners, poor but industrious, from Herzog, on the north side of the railroad tracks, bought their way in. They were Germans from the Volga region of Russia. Of Catholic faith, they (like the German Mennonites from the Volga region who settled in central Kansas to the south and east of post-rock country) came to the plains southeast of Hays to escape religious and political oppression. Variously called Volga Germans (generally preferred), Russian-Germans (preferred by some), and German-Russians (the most-used term), they were descendants of Germans who had migrated from Germany to Russia in the 1760s at the invitation of Catherine the Great, who had noted that they were an industrious, therefore desirable, people. Catherine granted them religious freedom and exemption from military service indefinitely (variously interpreted as one hundred or five hundred years or forever), as well as tax-free land for a certain time. In 1871, however, another Russian ruler, Alexander II, issued an edict limiting colonists' exemption from military service to ten years, during which time they could emigrate without forfeiture of property. That prompted these displaced Germans to send representatives to the United States and elsewhere to select, if possible, desirable places for new homes, according to the Rev. Francis S. Laing in "German-Russian Settlements in Ellis County, Kansas."[11]

Two of the investigators spent about a week in Kansas in the vicinity of Larned, but were not impressed with the land. Nonetheless, in 1875 a group set sail for America, debarking at Baltimore. Some agreement was made with the Atchison, Topeka & Santa Fe Railroad to bring the immigrants to Topeka. There many secured temporary employment while representatives, in the company of railroad officials, made several trips to central Kansas in search of a new home. Finally an agent of the Kansas Pacific showed several members some cheap ($2.00 to $2.50 an acre) railroad land in the vicinity of Hays. They sampled the soil and it "tasted after grain"—meaning they knew it was suitable for growing wheat and other grains, whether or not they knew it was rich silty loam. They decided they had found their new homeland.

Early in 1876 the first group of Volga Germans (or Russian-Germans or German-Russians) arrived in Hays, and almost immediately they moved south across the Ellis County line into Rush County and founded the town of Liebenthal. Also early that year twenty-three families landed at Victoria, first settling on the east bank of Victoria Creek, then moving to higher ground just north of Victoria to found Herzog, which evolved as the most important of the settlements. In the same year other groups arrived and laid out these Ellis County towns: Catherine, Pfeifer, and Munjor. In 1877 Schoenchen was settled by several families from the Liebenthal colony. The first homes of these Catholic Germans in the Kansas towns, named after towns they had founded in Russia, were board tents made with lumber hauled from Hays, but before the winter season they had built dugouts or sod houses. Being deeply religious, they built churches as soon as possible— temporary frame structures, then small stone churches, then larger stone churches to accommodate larger congregations.

Inclined to be clannish, the Russian-Germans had retained their German identity, including the German language, during their hundred years in Russia. Seemingly they were prepared to do likewise in Kansas, despite the isolating influence of homesteading and the so-called melting-pot atmosphere

House with back toward street, north Victoria. This stone house (owned by Albina Miller, Leona Russell, and Elsie Hoffman, granddaughters of the original owner) and a nearby one occupied by Mr. and Mrs. Celestine Sander are remnants of early Herzog.

of America. The communal character of their villages—houses built with backs toward the street and front doors opening on a rectangular courtyard—tended to unite the inhabitants socially. Slowly they became more public spirited and began to exercise more freedom in their manner of living, but even now many still speak German as well as or better than English, and a few of the square, native stone houses with a doorless back to the street may be seen.

A number of their forebears had been artisans when they emigrated to Russia, but most of those who came to Ellis County were farmers. A few even brought with them seed wheat, Turkish or spring. Though they suffered hardships that plagued other settlers, they were determined to become agriculturalists where others had failed. To obtain money to buy land and stock, or even to subsist when drought and other disasters struck, the men left the women in charge and sought work wherever they could get it. Many worked on the railroad west from Hays and beyond the Colorado line. Some hired out to the English colonists at Victoria. They collected buffalo bones, which could be sold for cash, traded for goods, or shipped east to be ground for fertilizer or to be made into harness ornaments or cutlery handles. Eventually they were rewarded for their perseverance, and ironically Herzog was incorporated into Victoria. Victoria—not the English cultural center it was to be

—today is very much an American–Russian-German town, and probably 75 percent of the farmers in Ellis County are of Russian-German descent.

Germans, of various denominations, have been influential in almost every post-rock community. While the Catholic Russian-Germans were locating in the vicinity of Hays, Protestant (predominantly Lutheran) Germans from the Volga were locating close by, to the east and south in an area extending from southwestern Russell and northwestern Barton counties into Rush and Ness counties. J. Neale Carman in "The Foreign Mark on Kansas" gives these loci: Milberger, Otis, Bison, and Bazine. Various groups spent a few years in South America, South Africa, or Mexico before reaching central Kansas. Additionally, as pointed out by post-rock country's notable judge and historian Jacob C. Ruppenthal, in "The German Element in Central Kansas," communities throughout post-rock country were receiving Germans of various faiths from Germany; from beyond the German Empire in Europe, including Bessarabia, Switzerland, Poland, and Austria; and from other states, especially Pennsylvania, but also Indiana, Illinois, Ohio, Missouri, Wisconsin, and Virginia. Among those who came in colonies or relatively large family groups were the Lutheran Germans from Bessarabia, who settled south of Russell in 1878 and 1879, and Pennsylvanian-Germans, including Quakers, Dunkards, River Brethren, and German Baptists, who migrated to various communities. Most who came from Germany came as family units or as individuals. Furthermore, most first settled in other states, later moving to central Kansas. Some tended to become part of compact settlements in Lincoln, Ellis, Mitchell, Osborne, Ellsworth, Russell, and Barton counties. Others simply moved to where they could obtain land or a livelihood. Thus, by 1880, Kansas had so sizable a German population that every county that had been organized had its German residents (census of 1880). If not by 1880, then not long afterward, almost every central Kansas school district had at least one German family.

Several sources note that of all the foreign influences in central Kansas, none exceeded that of the Germans. Despite grasshoppers, crop failures, and other adversities, most Germans held on.

STAYING ON

Swiftly turning the pages of regional history, we might be inclined to romanticize about staking out a claim in north-central Kansas, where once the buffalo roamed and there were no landmarks, not even trees, to interfere with the expanse of the prairies. However, we also become conscious of the courage and the resourcefulness it must have taken to stay on in this land of unpredictables and limited resources.

Imagine stopping your prairie schooner on a bit of prairie that you planned to call home. For temporary shelter, you might be obliged to live in your covered wagon, unless you had stopped near a creek bank where there was a dugout left by trappers or surveyors who had preceded you. For a time you probably could survive on wild game and edible plants, including perhaps a few plums and maybe berries. For water you would depend on springs. If you had arrived in a favorable year, say 1872, you might grow a crop by digging a few holes in the sod and planting a little corn. You probably would build a dugout or a sod shanty. You might even be optimistic about the future. But when in 1874 clouds of grasshoppers swooped down and devoured almost everything in sight, and then came drought and prairie fires, would you have stayed on, as did many pioneers? Or would you have added "ed" to the sign "Kansas or bust," which might still be legible on your wagon, and joined those who headed back east?

Staying on meant tilling the soil and developing an agriculture compatible with the vagaries of the plains climate. It meant finding and using natural resources to develop the region. The initial task, breaking the tough prairie sod, required special plows, which had to be made at home forges or local blacksmith shops. Plowing generally required a team of oxen; oxen could maintain their energy by eating range grass, whereas horses could not do heavy work without supplemental feed, which pioneers seldom could provide.

Surprisingly, many who came were not farmers; many who were had farmed in eastern United States, where tilling the soil was relatively easy and where rainfall generally was adequate for growing a variety of crops. The Volga Germans, who had farmed on the steppes of Russia, were cognizant of the type of farming that was necessary here, but most settlers were not. Furthermore, resources above the soil were limited. Timber was so scarce that substitutes had to be found for fuel, building, and fencing. For fuel, the pioneers used buffalo chips, compacted manure (*mist-holtz* or manure wood), dry grass, or lignite (a poor-grade coal found cropping out in many ravines).

Loose stones, picked up along the slopes, were found to be useful for dugout fronts as well as for dugout and cellar walls. Some pioneers even used loose stone as the principal material for houses and shelters. But initially the typical dwelling was of prairie sod, cut into blocks of about twelve inches by two inches by twenty or twenty-four inches and laid like brick to form walls twenty to twenty-four inches thick; the roof, of twigs and soil or sod, was placed over a central support called a ridge pole. Some pioneers mixed "native lime" (powdery crumblings from limy rock) with water to plaster the walls of their dugouts and soddies. Clay, which could be molded into brick and fired in kilns made of the raw brick, also was an early building material.

Though most settlers at first depended on springs for water, many dug wells as soon as they could find time. Not being geologists, they were guided

Dugout of stone on Cooper Gilt Edge Ranch, now a part of the Wilson Reservoir area, Russell County. *Photograph by Ralph Coffeen.*

by practical experience or sheer instinct (some perhaps having faith in the water-witching technique) in selecting well sites. It was not unusual for a homesteader to dig several wells, some perhaps as deep as eighty feet, without reaching water. Most who dug near streams in the valleys generally found water at shallow depths, thirty feet or less. For well linings, the pioneers used loose stones or pieces of rock broken from ledges of that "beautiful magnesian limestone" the promotional literature bragged about as being so abundant.

In getting a toehold on their prairie claims, settlers soon found that those ledges—those chalky limestones—were one of the region's most dependable and versatile resources. Even before the homesteading era, the native limestones in some way served anyone who frequented the area. Indians used broken pieces for burial mounds and fashioned crude weapons and implements from them. Explorers and trappers used them as markers. Surveyors made pillars of them for landmarks, and they partially buried limestone blocks to indicate section corners, scribing a cross at the appropriate spot on a block's exposed surface. Some of the early horse traders and cattlemen placed upended slabs side by side along ravine walls to form corrals.

But it was the settlers who came to stay who saw the most potential in the limestones and who in fact found them almost indispensable. Most used was the creamy-buff, brown-streaked layer that subsequently acquired the name Fencepost limestone, or post rock.

3
SAGA OF
THE STONE POSTS

Had it not been for stone fence posts, prosperity might have been a long time coming to much of north-central Kansas. Despite legislation to postpone or lessen the need for fences, settlers on the treeless upland soon found that fences were a requisite to area development. With native timber scarce and shipped-in lumber generally too costly for fencing material, the pioneers sought substitutes. When they found they could split posts from the rock layer that already was one of their main sources of building block, they added fence-post quarrying to their farming routine. Before the turn of the century miles of stone posts were holding up fence lines from Republic County to Ford County.

TO FENCE OR NOT TO FENCE

When the homesteading process began in central Kansas, most of the area was free range. That suited the cattlemen but not the incoming settlers who wanted to raise crops. By American custom the burden was on the crop grower to protect his fields from range animals. Farther east where timber for fencing material could be obtained readily, that responsibility generally was accepted with little complaint. But keeping range animals out of fields on the open prairie, where fencing materials were not readily available, was a different matter. Consequently, prairie settlers began to agitate for laws to shift the burden to the stockmen. The agitation first produced several ill-fated measures, including the herd laws of 1870 and 1871, which were declared to violate the state constitution because they applied only to certain sections of the state. Then came the herd law of 1872, which applied uniformly to all the state; individual counties, however, had the option of adopting it. That law was in effect in most central Kansas counties until it no longer was needed, sometime in the 1880s.[1]

Where in force, the herd law restrained livestock owners from allowing their animals to run at large and gave those suffering crop damage from livestock—whether or not fields were fenced—a lien on the animals until damages were paid. As a result, livestock owners began to herd their stock by day and tether them or keep them in corrals by night.

During approximately the same period, "any person planting an osage orange or hawthorn fence, or who shall build a stone fence the height of four and one-half feet, around any field, within ten years from the passage of this act, and successfully growing and cultivating the same, or keeping up said fence until it successfully resists stock, shall receive an annual bounty of two dollars for every forty rods [660 feet] so planted and cultivated or kept up; the bounty to commence as soon as the fence will entirely resist cattle, and to continue for eight years thereafter . . ." (*General Statutes of Kansas, 1868*, chapter 40, section 2, page 495). The county assessor was the judge of the merits of each fence, and the county commissioners were authorized "to order warrants to be issued upon the county treasurer."

The herd law and the bounty system, seemingly in conflict, in fact were complementary in that one intent of the herd law was to give prairie farmers time to build or grow fences. Biennial reports of the State Board of Agriculture help us feel the pulse of public sentiment on the herd law and its influence on fence building during the 1870s and 1880s. From reports for those years we learn that before 1880 the herd law was almost universally favored in central Kansas because "it stimulates the growth of small grains more than stock raising . . . [and] gives the settlers an opportunity to farm without the expense of fencing." One account noted as the only objection, "It gives the capitalist all the cattle, for the reason that parties with but a few head cannot afford to herd them."

The biennial reports variously interpreted the effect of the herd law on fence building:

> encourages hedge growing . . . its repeal would make hedge growing impossible, on account of the encroachment of cattle of the non-resident owners.
>
> tends to retard the growing of hedges and fence building.
>
> is not believed to have an influence upon fence building or hedge growing.
>
> retards hedge growing to a certain extent, but the farmers are planting it as fast as they become able.
>
> in favor generally with the people, yet very unpopular in some sections;

retards fencing and stock raising, but promotes general farming.

stimulates general farming; keeps stock safe until hedges are grown and fences built.

The problem of protecting crops from range cattle aside, public sentiment began to swing in favor of fencing as settlers acquired more stock. They too were obliged to herd their cattle, and herding was time consuming and costly, often requiring assistance of hired hands to supplement family effort. Before restrictions were placed on the cattle trade, in some areas resident stock in unfenced pastures commonly were exposed to the potentially fatal splenic fever brought in by Texas longhorns.[2] Furthermore, fencing seemed to be the only way to maintain quality herds by keeping range stock from mixing with resident stock. And confining animals to enclosed pastures, where shelter and supplemental feed could be provided, was a way to improve livestock. Additionally, fenced property tended to give the landowner a sense of security and well-being.

Motivated by such inducements, the prairie farmers began to take advantage of the bounty system. Whereas in 1874 the State Board of Agriculture reported "no fences" or "few fences" for most north-central Kansas counties, in 1878 it reported fences of all types—stone, rail, board, wire, hedge—for most of the counties. At first there were more rods of hedge than any other type. But by 1880 wire, for which no bounty was offered, had begun to prevail.

The bounty system notwithstanding, hedge fences proved not to be the answer to fencing in central Kansas. For one thing, it took from three to five years to grow hedge that would turn stock, even where climatic conditions were favorable. In the southwestern part of post-rock country, where "dry spells" were almost normal, establishing hedge was no small feat. The hedge seedlings that did establish a root system in the prairie sod as likely as not were consumed by wildfires or destroyed by other plains disasters before they matured. In the northern part, "hedges would grow, all right," according to Myron Chapman of Beloit, "but their roots would sap the soil moisture until crops wouldn't grow within 25 feet on either side of the hedge row." Furthermore, as pointed out by Bird Abram of Beloit, some stock would learn to "go through" the hedge.

Though some early settlers collected bounties for building stone-wall fences, that type of fencing on the prairies also had its drawbacks. Laying a wall four and a half feet high around large acreage took excessive time and skill even if that acreage provided all the rock needed.

Some homesteaders who had access to post timber began to build fences by using smooth wire, which had entered the market after the development of the telegraph in the 1830s. "The iron wire of that day," however, as pointed out by Earl W. Hayter ("Barbed Wire Fencing—A Prairie Inven-

Stone fence in northwestern Lincoln County (upper photograph); and remains of hedge fence, with row of full-grown hedge (Osage orange) in background, on the Peter Doctor homestead on the Scotch Plains, Republic County.

tion"), "was affected by temperature extremes; it snapped in cold weather and sagged in hot . . . had no terror for the livestock on the open range; they loosened the posts and broke the wire by constantly rubbing against it." For that reason, some individuals with inventive ability devised ways to attach barbs to smooth wire and to improve its tensile strength. Before 1870 at least three patents for barbed-wire fence had been registered at the U.S. Patent Office. It was not until after the Glidden and Haish patents of 1874, however, that barbed wire was marketed to any extent. Widespread use was delayed partly because of cost (Hayter says as high as twenty cents a pound in 1874) and partly because of damage inflicted on stock by the barbs, which were sharp and initially not rustproof.

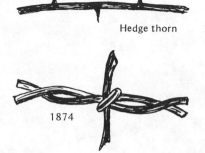

Hedge thorn

1874

One of the first barbed wires marketed (a Glidden patent of 1874) and a hedge thorn, which some say inspired the invention of barbed wire.

At the same time, somewhat paradoxically, wire emerged as a preferred fencing material. The Board of Agriculture biennial reports after 1880 noted that wire was used principally and barbed wire (legally sanctioned in the state in 1883) almost exclusively by 1890.[3] By then the cost of barbed wire had been reduced drastically, barbs had been shortened, and wire was of rustproof steel. As more farms were improved with fences, public sentiment turned against the herd law, which after the mid 1880s faded into oblivion from lack of need.

In prairie regions the lack of a source of fence posts initially restricted the use of wire, smooth or barbed, as much as did the wire's cost and quality. Growing timber for posts in much of central and western Kansas seems to have been no more successful than growing hedge fences. And those who took advantage of the Timber Culture Act of 1873, hoping to provide timber for building and fencing material, generally were disappointed. Under that act, in effect until 1891, 160 acres of land, in addition to the acreage offered under the Homestead Act, could be acquired by planting 40 acres— later 10 acres—of it to trees. The settlements and land acquisitions made under the act resulted in some valuable windbreaks but hardly a large timber supply.

Dim prospects of growing post timber and increasing awareness of the economics of wire fencing led pioneering farmers of north-central Kansas to the ledges of the hillsides for post substitutes. Old-timers of the post-rock region, old newspaper accounts, various histories (including the State Board of Agriculture reports), and other accounts of life on the central Kansas prairies confirm that by the mid 1880s stone posts were being used generally in north-central Kansas. It would be easy enough—and not far from the truth—to conclude that stone posts and barbed wire together emerged as the keys to development of that area.

For example, the *Sixth Biennial Report* (for 1887–88) said of Ness County, in the southwestern part: "A very superior hard magnesian limestone is found in large quantities in the southeast quarter of the county, especially near Bazine . . . in layers a few feet under the surface. It is easily quarried and prepared in desired shapes. The fencing in this part of the

Stone posts and barbed wire were keys to the development of north-central Kansas. Note wire attachment: staple plus nail driven into post. Vicinity of Victoria, eastern Ellis County.

county is mostly made with stone posts. Many miles of wire fence with stone posts have been constructed, and give the best of satisfaction. This stone withstands water and freezing, and hardens when exposed to the weather."

The *Ninth Biennial Report* (for 1893–94) added another dimension: "A hard limestone is found in great abundance in the southern part [of Osborne County], and makes excellent building material. It is also commonly used for fence posts, and is known in geology as post rock." That mention of "post rock" may be one of the first printed references to that term for the Fencepost limestone, a term that later became a regional trademark.

WHO SET THE FIRST STONE POSTS IN NORTH-CENTRAL KANSAS?

Some sources credit immigrants from Germany and Russia with quarrying and setting the first stone posts. Perhaps, but perhaps not. True, as we have noted, they recognized the potential of the native limestones for building material almost as soon as they arrived in north-central Kansas. But so did the British, the Bohemians, and the Scandinavians, not to mention American-born settlers who had been stoneworkers in eastern United States before arriving in Kansas. Quite probably individuals representing all those nationalities began quarrying rock for posts at about the same time in different localities.

We have found no documented evidence of the first use of post rock for posts. Even our chief informers, the old-timers, cannot say.

Articles in the centennial edition of the *Russell Daily News,* June 1, 1971, indicate that stone hitching posts and probably stone posts for corrals were in use in Russell County before the Volga Germans got to the Smoky Hill Valley. One article tells about an old stone corral on the Johnnie Schmitt farm on Big Creek south of Gorham:

> The corral, which must have been used as a camping area in the Civil War days by horse traders, was laboriously constructed by hand from rock. Shell rock taken from the nearby hills in large sheets was stood on end forming three sides of the corral. It stood as a solid fence six to eight feet high enclosing a sloping hillside area of about 200 square feet. . . .
>
> When [John] Roberts, a Welsh settler living in a dugout nearby, first visited the corral in 1872, he said it was falling down and showed no signs of recent use.
>
> The main gate for the corral was on the west where huge stone posts, standing over eight feet above the ground, had been erected.

The remains of those posts, the article stated, later were used in the spring well on the Schmitt farm and the limestone slabs in culverts and bridges in the area.

Mrs. Elizabeth Opdycke, Russell's senior citizen at the time of the centennial, recounted for the edition that she was five years old when she came to Russell in 1876 and that for three months her family lived in a building (in two rooms for $20.00 a month) on Main Street. Of her stay there she recalled "seeing a donkey hitched to a cart, standing in front of the house almost every day, tied to a stone post. The front door of our house was on the main street of Russell, and a stone or wood post stood in front of every building on either side of the street, to which horses, mules, donkeys and oxen could be tied."

Other than those references we found little to point to the use of stone posts before the mid 1870s and nothing to indicate widespread use before 1880. That is consistent with information on fencing in the early county reports of the State Board of Agriculture. In the 1870s most settlers were "just getting started." They had little livestock, so herding animals was no major burden. In addition, barbed wire had just entered the market, at a price few settlers could afford. Furthermore, quarrying rock for houses had a higher priority than did quarrying rock for fence posts.

Those who did quarry post rock for posts before 1880 surely were innovators. It seems pertinent, therefore, to name those about whom we have been able to secure information.

Adam Lee Faulhaber, who quarried tall posts in southwestern Mitchell County in 1875 or 1876, was one of the first of the stone-post quarrymen. A blacksmith, he was born in Germany and came to this country in 1866. He first settled near Minneapolis, Ottawa County (not far east of post-rock land), according to an unpublished genealogy of the Faulhaber family, made available to us by Donna (Mrs. Martin) DuVall of Salina and written by Mrs. DuVall's mother, the late Mrs. Allen Webster. Mrs. Webster wrote that Faulhaber, her grandfather, moved his family to the Round Springs community (east of Hunter) in southwestern Mitchell County in the spring of 1872. After digging a well, and finding water, he picked up loose rocks from the land surface and built a blacksmith shop—he had brought black-smith tools with him. Next he broke sod and planted crops; then for "almost a month grandfather was found in his shop making feathers and wedges for splitting out rock." The first stone quarried was used to make a large cave, which still existed in 1973, more than a hundred years later. He then used loose stones to build, adjoining the cave, a large room in which the family lived the first winter. Concerning her grandfather's practice as a blacksmith, Mrs. Webster pointed out that in 1875 and 1876 the families of Heller, Schneider, and Deiter, who had settled four or five miles west of her grand-father's farm, ". . . hired grandfather to quarry out tall stone posts. They were to be used in building livestock shelter and were the first stone posts he quarried. Each year found grandfather working a few more hours in the shop and making feathers and wedges that were then so much in demand."

Meanwhile, in the Bunker Hill vicinity in Russell County, about thirty miles southwest of the Faulhaber farm, a prominent limestone ledge there had become or was about to become a source of posts. An article in the *Russell Record* of February 10, 1900, stated:

> The idea of using [stone for] fence posts is supposed to have been conceived by C. F. Sawyer, who lived on the Saline river northeast of Bunker Hill. . . . A *Record* reporter talking to one of C. F. Sawyer's sons learned that Mr. Sawyer built this fence [with stone posts] in December 1878, and it is still standing and in use. Mr. Sawyer was a stone worker before coming to Kansas [from Illinois in 1873] and quickly recognized the possibility of using this stone for posts. He first used the wooden plugs for securing the barbed wire but later fastened it with smooth wire. He used feathers and wedges and had them made in Shaffer Brothers blacksmith shop in Bunker Hill.

The article clearly states that C. F. Sawyer set stone posts in 1878 but is not definite on when he first quarried posts—almost as soon as he "recognized the possibility" or later, perhaps not until 1878. Earle Sawyer, Charles Franklin Sawyer's grandson who operates a machine shop in Lucas, wrote recently that he is relatively certain the date was 1873 and that "the fence is still standing." It is highly unlikely that anyone quarried stone fence posts before that date, so Sawyer probably can be acclaimed the first quarrier of

C. F. Sawyer farm in the Elm Creek community in the Wilson Reservoir area northeast of Bunker Hill (or northwest of Dorrance), Russell County, as photographed by Leslie Halbe in the early 1900s. *From the Halbe Collection, Kansas State Historical Society.*

stone posts in the region. (It should be remembered, however, that although smooth wire probably could be obtained in 1873, barbed wire was not marketed to any extent until after 1874.)

Ralph Coffeen, of Russell, retired printer-artist interested in post-rock history, is convinced that stone posts were set around the Bunker Hill cemetery in 1876 and that that idea should be credited to a Scotsman, Joseph Menzies, who at the time was mining coal southwest of Bunker Hill. Coffeen, who perhaps has done as much as anyone to call attention to the area's post rock and diligently has checked information on its uses, stated (personal communication) that the 1876 date and the name Joseph Menzies both were reported to him by Clarence Peck, long-time maker of quarrying tools in Bunker Hill, and that the information was confirmed by Charles W. Shaffer, Russell banker and descendant of one of the Shaffer brothers, Bunker Hill's first blacksmiths.

Menzies, who had stopped in Pennsylvania before coming to Kansas in 1873, did not use feathers and wedges but "sledged" (or "broke out") the posts. Sledging, a common method of quarrying building block before the feather-and-wedge technique was employed extensively in the area, involved raising an uncovered slab with a crowbar or lever, propping it up with small pieces of rock, and breaking off various lengths with a sledge. During such an operation, a few blocks might have broken off in lengths long enough to give Menzies (and other resourceful pioneers) the idea of using them for posts.

Coffeen told us that he had heard but had no proof that men named James Chrudimsky and Charles Seirer in 1876 sledged out posts to use in a fence line on the Russell County side of the Russell-Ellsworth county border two miles northwest of Wilson. Verifying that hearsay took a circuitous route. In "Bohemians in Central Kansas," Francis Swehla notes that V. Chrudimsky filed a claim in that vicinity early in 1875, but that in January of 1878 Mr. and Mrs. Josef Veverka arrived from Chicago and "bought the relinquishments to the southwest quarter of section 2, township 14, range 11, in Plymouth township, Russell county, of Vac Chrudimsky. Mr. Veverka, being located on the ridge that divides the waters of the Smoky Hill and Saline rivers right where there are building-stone quarries in abundance, made good use of them, putting up all farm buildings, and even a corral, of the magnesia limestone."

The Register of Deeds office, Russell County, gives 1884 as the date that James Chrudimsky was issued a patent on the southeast quarter of section 2, township 14, range 11, which would adjoin the quarter section bought by the Veverkas. Personal communication with area residents led to correspondence with a granddaughter of James, Mrs. Emil Zahradnik of Wilson, who verified that James and Vac were the same person (Vac, for Vaclav, translated from Bohemian to English, is James). A lucky contact was that with Madge Seirer Hooper of Sylvan Grove, who referred us to Parker V. Seirer of WaKeeney, who established that his father, Charles Seirer, purchased and lived on land (section 1, township 14, range 11) that adjoined Chrudimsky's. The property, two miles northwest of Wilson in Russell County, borders Ellsworth County on the east. Parker V. Seirer also verified the statement that his father and neighbor James Chrudimsky sledged out posts and set them on the Seirer property in 1876. Some of those posts are still standing, he said. Later, his father and another neighbor, Gus Deissroth, used feathers and wedges to quarry posts, and they provided nearly all the slabs for sidewalks and crossings in Wilson.

The Volga Germans who settled in Rush and Ellis counties south and east of Hays between 1876 and 1879 turned their attention to quarrying rock for posts after first quarrying it for houses. In the Liebenthal vicinity, where most sources say the first group stopped, that was about 1878, according to J. M. (Jake) Herrman, Liebenthal's oldest living resident in 1973. Relying on information he recorded after conversations with his grandfather and father, he said he knew of no precedent for stone posts, only the pioneer's ability to make use of available material.

The Volga-German settlements of Schoenchen and Pfeifer in southern Ellis County north of Liebenthal probably had stone posts about as early as did Liebenthal. Mrs. Charles Hill of Wilson said that her grandfather Breit, one of the original settlers at Pfeifer, often told her about the unique means of quarrying rock for posts.

Stone posts evidently were being used to improve at least a few homesteads in eastern Ness County even before that county was organized. In one of her earliest reports (February 14, 1879) to the *Plattsburgh* (New York) *Republican,* Mrs. Seymour Mooney, a member of one of the first families in the Bazine community, wrote: "Speaking of stone houses reminds us of the quarries which furnish plenty of building material. The stone is easily dressed out and of nice quality. They lie on and near the surface and also form a dividing line between the bluffs and ravines. Stone are used for fence posts and are got out as easily as those of wood. One gentleman has six hundred acres fenced, using stone posts and barbed wire."[4]

Mrs. Mooney did not identify the gentleman, but a good guess is that he was John H. Farnsworth, evidently the first person to prove up a claim in Ness County and in so doing to use stone for building material and other uses. On his property, along Walnut Creek and including the present site of Bazine, limestone was plentiful.

About 1880 a Swede, who had been a stonemason but who then was working on the James Webster homestead in northwestern Lincoln County, was asked to carve a grave marker of native limestone for a resident who had died. According to correspondence in 1961 between Mrs. Allen Webster and C. R. Hubbard of the Guaranty State Bank of Beloit, the Swede had noted while carving the limestone that it was easy to work. Keenly aware that material for fence posts was scarce, he offered to make posts of the rock to fence the quarter section of land on which the Websters lived. Presumably he then took on the task of quarrying and shaping stone posts during the slack winter months. As a result, in that vicinity the Webster property became an exhibit of the use of stone posts in fence lines. Mrs. Webster noted, however, that all the property was not fenced until about 1885.

For the initial fencing on the Webster place, smooth wire was used; it was not replaced by barbed wire until the 1890s. That may have been because smooth wire was cheaper than barbed, or because barbed wire was not legally sanctioned in Kansas until 1883, or because the Websters were cautious—aware perhaps that sharp, rusty barbs could "ruin a lot of stock."

Harold Dwyer, veteran newspaperman now with the *Hastings* (Nebraska) *Tribune* and formerly of eastern Mitchell County (Kansas), believes that one of the first uses of the Fencepost layer for posts, at least in the northern half of post-rock country, was on the T. H. Oakley homestead on the divide north of Asherville, one of the first permanent settlements in the Solomon Valley. The Oakley homestead was patented in 1880, according to records in the Mitchell County Courthouse. If the Oakleys lived on the homestead for five years before the patent was issued, they could have included fences with stone posts in their improvements in those five years. After talking with the widow of one of T. H. Oakley's sons, Mrs. Bert Oakley of Asherville, we decided that Oakley probably didn't begin setting stone posts until sometime in the 1880s. Mrs. Oakley remembered that near the turn of the century he was quarrying many posts to sell, usually hauling them to their destinations.

Bird Abram, retired blacksmith and stonemason of Beloit, believes, as does Dwyer, that the first rock quarried in Mitchell County was in the Scottsville and Asherville vicinities. He noted that his father, C. W. Abram, and Samuel Carter, who farmed between Asherville and Scottsville, were among the first in the county to sledge out rock for cattle sheds and barns. Also in the 1870s some Scotsmen and Scandinavians from the settlements in southwestern Republic County began to take up claims along this divide, where "they could find plenty of rock for building." Inasmuch as many of these people were stonemasons, their knowledge of rock and their inclination to use it suggest that in this area quarrying stone posts would be a natural extension of quarrying building block.

Even though we cannot determine who first quarried and set stone posts—and where and when—in post-rock country, it is certain that by the early 1880s the idea had been introduced in most parts of the region. On the other hand, the practice did not become well established until the mid to late 1880s. By then many settlers had added to their livestock until herding had become a burden. So as the price of barbed wire came within their reach and their housing needs had been met, they began to devote time to quarrying rock for fence posts.

Our search indicates that the State Board of Agriculture's first published mention of the use of limestone for fence posts appears in the *Fourth Biennial Report* (1883–84). This entry is for Mitchell County: ". . . limestone abounds in nearly every portion, is of good quality, . . . is used largely in making fence posts."

Soon stone posts had become so common in the area of the post rock's occurrence that members of the first generation born in the region accepted their use as routine. For example, Margaret Haun Raser, who was the first child born (1881) in Jetmore, in 1960 said of the Hodgeman County homestead where she spent her childhood, "As early as I can remember the pasture was fenced with stone posts."

Thomas Frusher, who was two years old when his parents brought him from England to Hodgeman County in 1878, early learned the art of quarrying posts and later became one of the best stonemasons in the Jetmore vicinity. In a letter written in 1962, he gave 1883 as a probable beginning date for the setting of posts in that area, and he succinctly stated the "why" of the extensive use of stone posts: "They [stone posts] were utilized in the area because they were the material at hand on the prairie. Western Kansas was destitute of timber at that time. Another factor favoring their use was . . . they did not have to be replaced when the prairie fires swept across the country. Sometimes these fires came for a hundred miles or more and the old stone fence post was the one thing that could withstand them."

HOW THEY DID IT

Most of us can imagine how a pioneer farmer might go about securing and setting wooden posts. And few of us would doubt that one man alone probably could handle much of the work involved. But try to imagine the task ahead of a farmer who, for want of post timber, decided to split posts out of rock strata on his land. Such was the choice and the task of many a north-central Kansas pioneer whose homestead was blessed not with trees but with conspicuous ledges of chalky limestone.

The chalky limestone, however, had advantages. It was close enough to the surface (directly under the topsoil in much of the region) or was well enough exposed in ravines or along hill slopes that it could be obtained easily with the proper quarrying tools and techniques. It was uniform, seldom less than eight inches or more than twelve inches thick. It was persistent, extending with little interruption for many miles. And it was soft enough to shape with chisel or other tool when freshly quarried but would harden after being exposed to the air.

But there were disadvantages. Besides the skills involved, quarrying rock in "post" lengths was hard work and time consuming. Once split out and shaped, the posts had to be hauled to the fence line. That meant more hard work, as well as ingenuity on the part of the farmer or quarryman, in that a rock post five or six feet long and about eight by ten inches in cross section weighed 350 to 400 pounds.

Accustomed to hard work, resourceful out of necessity, and determined to have fences, the pioneers in post-rock country developed workable techniques for quarrying, shaping, and setting stone posts. It probably helped that nearly every community had at least a few stoneworkers ready to use their skills or willing to give advice or to demonstrate "how it could be done."

Whereas some early settlers used a sledge to break out building block from rock ledges, using that method to produce blocks long enough for posts would have been impractical, if not next to impossible. True, there is some evidence that a few quarrymen achieved the impossible. For example, if we examine the stone posts around the Bunker Hill cemetery, we can find a few that have no drill marks, indicating they were sledged out; quite likely they are survivors among those set by Joseph Menzies in 1876. While acknowledging Menzies's triumph, we can state categorically that reliance on sledging would have resulted in few stone posts for north-central Kansas.

The feather-and-wedge method (sometimes referred to as the feather-and-plug or, rarely, as the feather-and-wing method) made post quarrying feasible. It is not clear who developed the process. Quarrying, however, is an ancient art, probably used in some form in various parts of the world long before C. F. Sawyer, Adam Lee Faulhaber, the Volga Germans, and their contemporaries in post-rock country were adapting the process to split out posts. Wayne Barnett, construction contractor of Glen Elder, suggested that feathers and wedges were known to early Egyptians, whose slaves reportedly used them to quarry rock for the Pyramids. Indeed, several references in encyclopedias credit the ancients for the process. Harry Grass, banker at La Crosse and stone-post enthusiast, pointed out that, in more modern times, an Australian who visited the Post Rock Museum knew immediately the meaning of the term "feather" as applied to rock quarrying.

Bird Abram reminded us that in Kansas drill marks may be found on stone buildings constructed more than a hundred years ago, before the quarrying of stone posts, and Harold Dwyer observed that feathers and wedges have been used to quarry building stone from other types of rock in other Kansas areas. Further, *Webster's Third International Dictionary* gives this definition for feather: "One of two wedge-shaped short metal rods curved at the upper end and driven into a hole drilled in rock and forced apart by another rod driven in between them in order to split the rock."

Seemingly, then, nothing was new about quarrying with feathers and wedges. Be that as it may, Clarence Peck of Bunker Hill called the process an improvement over sledging and "in the better farming interests." He emphasized that the "plug-and-feather" innovation was of considerable importance in the development of the area and "made possible the fencing of many cattle ranches: Rockefeller, Cooper, Anspaugh, Shaffer, Johnson; the Britt horse ranch, to mention a few."

After the feather-and-wedge method was adopted in the area, it was used almost exclusively not only for quarrying posts but also for quarrying building stone. In fact, rock that broke in lengths too short to use as posts generally was set aside for building block.

Tools used in the quarrying and shaping process were simple. Besides feathers and wedges (plugs), they included: stone drills and bits of various types, chisels, stone hammers, slips and scrapers, and scribers. Most of the tools were made at home forges or in local blacksmith shops. A blacksmith in the pre-automotive era was a community necessity, and in post-rock country nearly every blacksmith was called upon to make quarrying equipment. Some became adept at it, if not specialists. Peck recalled, for example, working for Charles Shaffer, a "superior operator and maker of tools when that community [Bunker Hill] was young." Peck himself became known in the Bunker Hill area as the "best maker of bits for hand drills used in quarrying." He also was skilled in making plugs and feathers.

A number of pioneers, having no money to buy tools or to have them made, improvised. As an example, Coffeen mentioned that feathers might be made of metal strips from wagon boxes and wedges might be railroad spikes or even oak plugs. A drill bit could be made simply by heating a metal rod, or preferably a large worn file, and twisting it.

To ease the drilling task, some quarrymen used a drilling buck—a saw-horse with a long drill attached to one end.

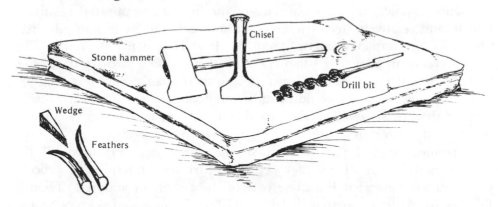

Some tools used in quarrying post rock.

Exposures of the Fencepost bed near hilltops were potential quarry sites. To open a quarry, as described by Harold Dwyer in a recent letter, "the first step of course was to discover the outcropping of the proper thickness and location. Then plow the soil overlay and remove it. This was done with a horse-drawn scraper commonly known as a 'slip.' One would hold the handles and fill the thing as the horses pulled it along. When it had been taken to where it was to be piled, a lift on one handle would catch the point in the soil and the pull of the team would dump it—and be ready to return for another load."

The overlying material ranged from a few inches to three, perhaps four, feet thick, though generally it was considered impractical to uncover rock lying much more than three feet below the surface. Jake Herrman said that in the Liebenthal vicinity up to three or four feet of dirt might be removed, depending on how many posts were needed and how large a slab was to be uncovered. "The farther back you had to go to get these rocks out the more dirt you had to take out," he said. Thomas Frusher, in a letter written in 1962, said that in Hodgeman County the depth of the overburden ranged from a "few inches to several feet." According to various sources, a slip used to remove overburden had a load capacity ranging from three and a half to five cubic feet. Sometimes a larger scraper, with one centrally placed handle, was used. Called a fresno scraper because it was made in Fresno, California, it could scoop up to a cubic yard of overburden at a time, according to Bird Abram of Beloit (Wayne Barnett of Glen Elder thought generally less than a cubic yard). Commonly it took two teams of horses to operate a fresno, and when the operator raised the handle to dump the load, he had to know what to do to keep from being dumped also, Abram and Barnett said. Pioneers who did not own a team simply used hand power to uncover a slab of rock. Samuel Carter, early quarryman of Scottsville, used a shovel to remove the overburden.

Once exposed, the upper surface of the limestone appeared relatively smooth and relatively free of joints—"unbroken for many feet in all directions," as one former quarryman put it. Jake Herrman observed: "The rocks were all in a layer but had cracks in them. Some of these slabs were probably ten, twelve, fourteen feet wide and more than twenty feet long . . . in some places the rocks would be only four feet square." Thomas Frusher claimed that the slabs uncovered varied from four or five feet to as much as twenty or thirty feet in length.

Having removed the overburden, the quarrymen measured off the lengths of the posts. Then they began the drilling and splitting process, which varied somewhat from one part of the region to another. Thomas Frusher in his letter written in 1962 told how he quarried posts near Jetmore:

> The only tools used for the actual splitting of the posts [were] a wood auger with the center cut out, feathers and wedges and about a 4 lb. hammer. The feathers [and] wedges . . . were driven into the drilled holes . . . small end of feathers up, large end down in the drilled hole. Then the wedge was driven small end down between the feathers. This caused the rock to split before the small end reached the bottom of the hole. The holes for the feathers were drilled about 8 inches or 10 inches apart in a straight line.

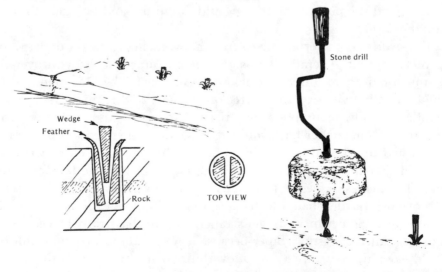

Feathers and wedge in place (from sketch by Clarence B. Mehl) and a stone drill (sketch of one donated to Post Rock Museum by Ralph Coffeen).

Harold Dwyer in a recent communication has vividly recorded how it was done in Mitchell and Jewell counties in the early 1900s:

With the topsoil removed, a straight-edge was used to help outline and mark the posts the proper length and width on the surface of the stone. Holes were drilled some eight inches or a foot apart along these lines. This was done with a home-made drill . . . the drill . . . is really an elongated "brace and bit." The bit is a wood bit, as I recall about a ¾ or an inch . . . welded to a grain wagon endgate rod which had been to the blacksmith shop to be shaped to make the "hand-hold" near the top at about waist height. To this has been added the fly wheel off of a hand cornsheller just above the bit to enable the operator to keep up momentum as he drills into the stone. The "tit" on the top of the bit which sucks it into the wood as pressure is applied and the thing turned, was ground off for the stone work, it not being necessary. The holes were drilled about ⅔ through the stone, then the feathers and wedges put in.

After the posts were posts, they didn't exactly jump around in the quarry under their own power. To get them up where they could be handled, it was the custom to take a spadeful of dirt from under one side, making a hole large enough to receive the end of a log about 15 feet long, under which a smaller stone could be placed to "pry" over with the necessary manpower pulling down on the end of the log. The tools were simple, but they worked if given the required amount of elbow-grease. And you may be certain that after a day in the quarry one didn't need 18, or even nine, holes of golf to be ready for bed.

The feathers used in breaking the stone posts loose from the stratum were of metal, rounded on one side and flat on the other, obviously so one side would fit the side of the hole in the stone, and the other side fit the flat side of the wedge as it was driven between the feathers. . . .

The wedges were tapped gently on top with a stone hammer, alternately from one end of the post to the other after the feathers and wedges had been placed in all the holes on that side of the post.

There was a warning sound when the pressure of the wedge had started the stone to crack. This "ping" was the word to NOT tap that wedge the next time past, but to tap the others in turn until they too had "pinged." When all had—it was a post.

Charles F. Speck of Sterling (formerly of Hoisington) has provided this account on quarrying posts in Barton County:

We would take a team and slip and uncover the rock. Then we would bore 3 holes, one on each end and one in the middle. The holes on the ends were about 6 to 8 inches from the ends. Then [we would] have wedges to put into the holes. The one in the middle would take about 2 taps more than the wedges in the ends would take. The drill bit would be ½ inch to ¾ inch, whichever one Dad wanted to use. . . . The finished posts would be about 6 feet long and about 8 or 10 inches square on the ends. . . . The posts should be cut out as soon as the rock is uncovered, while still soft, as the sun and air will quickly dry and harden the rock.

Clarence B. Mehl, Jr., who lives south of Beloit, Mitchell County, and is one of the few persons still quarrying building stone on a somewhat regular basis, explained that after scraping off the overburden, his father used a crowbar to find "blind seams"—incipient cracks in the rock formed by unequal pressure or other natural processes. Having determined the blind seams, he marked off the line for drilling parallel with the seams so that when he split out the posts they wouldn't break in the wrong places.

Harry Grass, La Crosse banker, though unfamiliar with the term "blind seams," said that quarrymen he had talked with looked for seams resulting from natural stresses and strains, because the rock would break along those seams and "if there was not enough length between seams they would end up with building stone rather than fence posts."

According to some sources, a few farmers secured posts during the winter by removing the overburden above the Fencepost bed, drilling holes in the rock, and before a freeze pouring water in the holes. Water freezing in the holes is said to have cracked off the posts. Other sources doubted that such a method was ever used. Harry Grass, whose interest in the Post Rock Museum has led him to make many inquiries in search of authentic information on post rock, said that he has not yet been able to verify the freezing method. He noted that one practical problem in using the method would be to get water in the holes shortly before a freeze; otherwise, the water surely would soak into the rock or would evaporate.

Clarence Youse of Dodge City, formerly of Bunker Hill, reasoned that because the soil above the rock usually was removed when the ground was not frozen, it might be difficult to shape posts out of a rock layer that thus could have been hardened by prolonged exposure to the air. Golden Morris of Lincoln said that he had split out many posts over several decades but had never heard of freezing them out. However, other reports indicate at least limited use of the freezing method to quarry rocks.

Floyd Duncan, recently retired from the staff of the physical plant at Kansas State University but formerly a resident of Russell County, vouched for the procedure. He noted that during the World War I years and in the 1920s he helped his father split out rocks by that method. Duncan explained, however, that holes drilled in the rocks had to be larger when the freezing method was used than when the feather-and-wedge method was used, and that the water had to be poured in just before the freeze. He intimated that the method was not practical.

Duncan said that when he and his father used feathers and wedges, they commonly first uncovered the rock layer and let it freeze before drilling, because rock would split easier when frozen. Harvey Roush of Lincoln mentioned that his father also preferred to quarry rock when it was frozen—for the same reason.

Coffeen remembers that Billy Gaines, a stoneworker employed on the Cooper Gilt Edge Ranch north of Dorrance, told him that he once quarried rock by allowing water-soaked plugs to freeze in drilled holes; he used the method simply because "there was no money to buy material to make feathers and wedges." A few sources have indicated that the "soaked-plug" method could be used also in the summer; when the soaked plugs expanded in the drilled holes, they split the rock.

The "softness" of the freshly quarried posts varied with the depth and amount of moisture in the ground. The posts could be shaped easily with drill and hammer while still soft; that is, before they were hardened by prolonged exposure to air.[5]

Once cut out and shaped, the stone posts, according to most sources, were allowed to "season" before they were placed into the post holes. Some old-timers, however, could not recall a deliberate waiting period. C. H. Scholer, who grew up near Barnard, Lincoln County, and later became head of the Department of Applied Mechanics at Kansas State University, in 1962 said that he generally did not make a point of seasoning. Margaret Haun Raser, who helped her brother quarry rock in Hodgeman County, did not remember a "drying" period.

Stone post in fence line along country road southwest of Barnard, Lincoln County.

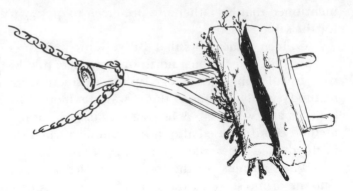

Forked tree limb used for sled to haul stone posts, as described by Ida Brown.

The posts were delivered by various means to the field or pasture to be fenced. Scholer, who claimed that at age seventeen he could lift and handle alone a seven-foot post (weighing at least four hundred pounds), attached posts to chains so that horses could pull them to the post holes. Mrs. Ida Brown of Lincoln said that she helped her father, Karl Kanzius, in the quarries on their farm in northwestern Lincoln County. To haul the posts short distances, Kanzius—who, incidentally, was born in Potsdam, Germany, and came to Kansas in 1877—and helpers used a "sled" or "boat," which Mrs. Brown described as a "large forked tree limb" on which they laid branches crosswise over the forked part to make a platform. Then, after managing to get several stone posts on that platform, they put a chain around it, hitched it to a team of horses, and pulled the sled to the post hole. (Mrs. Allen Webster described a similar conveyance called a lizard. "Wishbone" was another name for such a sled.) Mrs. Brown said that if the fence line was a mile or more away, several men lifted the quarried posts onto a lumber wagon to haul them to that destination.

Getting the posts on the wagon bed—or on any other conveyance—was no job for the weak. Golden Morris placed a chain under the posts, to lift them to the wagon. Jake Herrman said: "I used to haul some pretty heavy posts, and I used to take them all by myself and set them up. We would load them on a wagon. We had to lift them on. I would drag the post up [to the wagon bed] and lift one end up and put a roller—a pipe or whatever I had—under it and roll it on the wagon. Ten posts made a pretty good load."

Floyd Duncan mentioned that he knew one quarryman who, when he had a pile of posts ready, would dig into the quarry slope and back his wagon into the slope so that he could push the posts on the wagon bed.

The block-and-tackle method was used to load posts on wagon beds at some quarries.

Once the posts were delivered to the fence line, it was a simple job, according to Duncan (and others who had become inured to the routine), to tip the posts into the prepared holes. The heavier end of the post, always, was dropped into the hole. Depending on the height of the posts, the holes had been dug by hand—at first using a spade, later an auger or post-hole digger—to a depth of eighteen inches to two or more feet. Holes were dug at various distances apart, commonly about every thirty feet, so that in the finished fence line there would be from 160 to 175 posts to the mile. According to Bird Abram, some farmers set posts as close together as ten to fifteen feet. Ralph Coffeen quoted old-time quarrymen as saying simply, "about ten steps or more apart."

Anyone traveling through the stone-post area can observe how corner posts were propped to stay in a vertical position: by leaning other posts against them at about a forty-five-degree angle, generally in the direction of the fence lines.

Corner post in vicinity of Barnard, Lincoln County. *Photographed by C. C. Abercrombie, 1956.*

Posts were prepared for the fence wire in several ways. Perhaps one of the most popular and successful was to notch the post's edges at appropriate spacings for the strands of barbed wire, so that smooth wire could be wrapped around the post and the ends twisted around the barbed wire to hold it in place. Some farmers, Coffeen said, used an old carpenter's saw to cut a shallow notch in the post face to keep the wire in place, and, "As a rule of thumb, the lowest barbed wire was fastened about a carpenter's hammer length above the ground and a second wire the same distance above that and so on." Three to eight strands of wire might be used—"three to

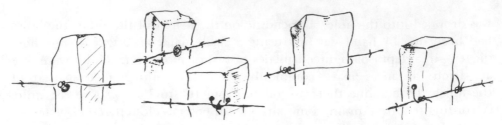

Wires were attached to stone posts in various ways.

hold cattle or horses, five for calves and sheep, seven or eight for hogs," according to Bird Abram.

Wayne Barnett of Glen Elder noted that the wrap-around method had a major advantage, assuming the posts were set properly; that is, aligned in the fence line so that the split sides (in which the relatively hard brown streaks showed) always faced in the direction of the fence line. When changing temperatures put stress on the wire it did not break but rather cut into the notches. He explained, however, that if the posts had been set so that the barbed wire stretched over the split face of the posts, then the wire would be subject to breaking, because it would be unable to cut into that relatively hard inner surface of the post layer.

We had often wondered why we seldom saw a fence post set so that its brown streak faced the roadway!

Drilling holes in the corners of posts provided another effective way to attach strands of wire. When posts so prepared were set in the fence line, smooth wire would be looped through the holes and twisted around the barbed wire. That method, though perhaps a little more time consuming than the wrap-around method, was preferred by some of the most meticulous workers. Arthur Jepsen, for example, used only that method, claiming that "attaching the barbed wire to smooth wire wrapped around the posts was a lazy man's way of doing it." Opinion aside, exhibits of both methods may be seen in fence lines throughout post-rock country.

One of the first, but somewhat unsuccessful, ways of preparing stone posts for wire attachment was to drill holes in the faces of the posts, drive wooden plugs in the holes, and fasten the wire to the plugs with staples. In "The Mudge Ranch" (written for the Hodgeman County Historical Society in 1931 and later published by the Kansas State Historical Society), Margaret Evans Caldwell indicates that method was used on that famous Hodgeman County ranch established in the early 1880s. Coffeen, in evaluating the method, reported in the *Russell Record* in 1941: "At first smooth wire was scarce and three or four holes were bored in the face of the post the distance apart the wires were to be spaced and wood plugs driven into the holes. The barbed wire was placed over the plug and secured with several nails or staples. The use of the plugs was not satisfactory because they worked loose and the wires would fall from the posts."

LAND OF THE POST ROCK

Here and there stone posts with bored holes in their faces are still standing. Some even contain remnants of the plugs. Good exhibits may be seen in the vicinities of Bunker Hill, Wilson, Sylvan Grove, Liebenthal, Schoenchen, Pfeifer, and Walker.

Posts prepared for special fences—around yards, cemeteries, or other places where decoration was a factor—might be tapered or otherwise shaped, perhaps even carved, and holes might be drilled so that rods or chains could be extended from post to post. Examples of special fences may be "discovered" by driving about in post-rock towns.

In one or two places we observed double rows of fences, perhaps four or more feet apart. We pondered about that one until Coffeen told us that they were "spite" fences, set up by neighboring farmers who were unsure of their boundaries and wanted to avoid legal battles. Harvey Roush referred to the space between as "the devil's lane." He added that many double fences were temporary, in which event stone posts were not used. On the

Spite fences, along US 183 between La Crosse and Liebenthal, Rush County (upper photograph), and along Kansas 14 north of Lincoln, Lincoln County.

other hand, use of the hard-to-install posts for such fences provides some insight about the stern individuality of these pioneers. Others used a second fence to discourage stock in adjacent pastures from breaking through for visits that might result in mixed-breed offspring. In addition, some farmers were not sure that their neighbors would do their share of keeping up a jointly owned fence. In a mimeographed pamphlet, "Fencing: Facts and Fancies," George Jelinek of Ellsworth points out that when horses provided most of the farm power and many horses were kept in pastures, the horses would rub noses across common fences and paw at one another, often fighting, until they caught their feet in the barbed wire, resulting in badly cut legs or occasionally injuries that dictated death to stop the animals' suffering. Double fences kept them apart and uninjured.

OLD-TIMERS RECALL THE QUARRYING BUSINESS AND THE WORTH OF A STONE POST

How many hours a day, one might ask of old-timers, were spent in the quarry and for what wages? Without hesitating, many would reply, as did Jake Herrman, "about fourteen hours, for $1.00 to $1.50 if you did it for wages." That would have been in the late 1800s and early 1900s.

Speaking for old-timers of his acquaintance, Harry Grass estimated that "if the overburden had been scraped off, a good man probably could break out about two posts an hour." Many old-timers recall that breaking out twenty to twenty-four posts was "a good day's work." Sarah Peters, in an article appearing in the *Kansas City Star* February 8, 1955, recorded what Louis M. Cooke of Topeka remembered about a one-time family project to fence 400 acres in Mitchell County: "After . . . the rock layer lay exposed and was 'squared up,' a 'working gang' could get perhaps a hundred posts a day. . . . This, however, meant a 16- or 18-hour day of back-breaking labor."

Not all farmers in post-rock country had accessible outcroppings of post rock on their property, or, if they had, for some reason they could not or chose not to go into the quarrying routine. To get their posts, they bought them from, or did some trading with, their neighbors. For a nominal fee some permitted quarrymen to split out posts on their property, and "once a quarry was opened the landowner usually would use it," according to Floyd Duncan. In 1941 Coffeen gleaned enough information from the old-timers to write in the *Russell Record* that "a number of the pioneer farmers would work together quarrying the posts and to many of them it was an all winter's job after the wheat drilling, butchering, and putting up feed was over. Some men made a year around business of it."

The men who made a business of it probably included both those who worked in quarries for wages and the farmers who saw the rock ledges on their land as a source of needed cash. They might work seasonally or when

Quarry near Olmitz, western Barton County, in the 1920s. Note various lengths for fence posts and building block (pile of building block in background). *From* Geology of Ellis County *by N. W. Bass.*

they could spare the time from other farm labor. Or, especially if the family was large enough to enforce a "division of labor" policy or if the available rock was extensive enough to warrant the use of hired labor, they might keep the quarries open from January through December.

Paul Adams, formerly of central Kansas and now an independent geologist in Oklahoma, in a letter several years ago gave this light touch to the turn of events for his grandfather, John Altenbaumer, a blacksmith who declined an offer to be deputy sheriff of Ellsworth: "My grandmother didn't like the sheriff business so she got the idea to purchase the farm which was located near Ellsworth. He opened a limestone quarry on the farm and so, instead of becoming a sheriff, probably famous (and maybe dead), my grandfather became a limestone quarrier. For a happy ending, the farm was paid for with fenceposts and building block from the quarry."

Mrs. Bert Oakley revealed that her father-in-law, T. H. Oakley, who had "a lot of rock" on his land north of Asherville, "quarried for all of his neighbors and hauled many loads of posts." Oakley must have had superior posts or salesman ability. Harold Dwyer recalled buying one 500-pound post from him in about 1903 for fifty cents, a better-than-average price.

Some of those who sold posts from their quarries didn't always display their product to the best advantage on their own land. Harry Grass tells this story: "A person with whom I am well acquainted asked his father why the neighbors all had such straight, pretty posts and they had all crooked posts that were not so attractive. His father's reply was that he sold the straight ones and 'planted' the bad ones on his own property. Thus, the man who was in the business of marketing posts would sell the choice ones

Stone post near Olmitz, Barton County. Note the off-center streaks. (Some variations in brown streaks on stone posts in post-rock country shown in four-page insert of color plates.)

and use the others in his own fence lines. Broken posts he would dress and use for building block."

And what might be a fair price for a stone post? Anyone who could find his way to the quarries on the Clarence Mehl, Sr., farm south of Beloit in the early 1900s could buy a stone post for five cents. "That's what Dad said he sold posts for at that time," said Clarence Mehl, Jr., who has reopened one of the quarries—not to obtain posts (except for his own use) but to obtain building block.

Reportedly, posts could be bought for five cents each in some other parts of post-rock country, also. In at least one locality, stone posts once were marketed for three cents apiece. That is the price M. K. Mathews of Quinter said his father received for posts quarried in the vicinity of Luray in 1892. Ruth (Mrs. Leland) Shaffer of Bunker Hill said that at about the same time her grandfather, James Loper, who lived seven miles east of Luray and six and a half miles northwest of Lucas, quarried and set posts for three cents each. Her father and an uncle—Charles and Luther Loper—helped him, though they were quite young. Ten cents a post seemed not to be an unusual price in the 1890s, however. During that decade Joseph Riedl, Austrian-born father of Mrs. Monte Krug of Great Bend and grandfather of Mrs. Russell Townsley of Russell, quarried and sold posts for that price in northwestern Barton County.

During the Spanish-American War, Paask Johan (P. J.) Jorgensen, who came to Lincoln from Denmark in 1887, was splitting out about twenty posts a day and selling them for ten cents apiece, according to his son, Hans J. Jorgensen of Lincoln. (Mrs. W. Carl Johnson, Salina, whose family was acquainted with this post-splitting Dane, said that he "came to be known as Post Jorgensen.") In the early 1900s, when he was a teen-ager, Emil Mauth of Bazine sold posts for ten cents each, also.

Though prices varied with time, during most of the era of post quarrying, posts were worth somewhat more than ten cents each. In Rush County the *Walnut City* [Rush Center] *Herald* of March 17, 1886, reported that stone posts sold for as much as twenty-five cents each. Jake Herrman said that at Liebenthal he paid twenty-five cents a post in 1914 but thought that his father earlier had bought posts for less. Clarence Peck, C. H. Scholer, and Clarence Youse each thought that during the two decades before 1920, twenty cents a post at the quarry, or twenty-five cents delivered, was a good price, at least in Russell and Lincoln counties. Golden Morris remembered quarrying, setting posts, and attaching the wire to them for fifty cents a post in the early 1900s. Erasmus Haworth, long-time state geologist, placed the market value of stone posts at twenty-five cents each, delivered to the fence line from quarries as far as fifteen miles away (*Kansas Mineral Resources for 1900 and 1901*).

Thomas Frusher, who considered the useful period of quarrying post rock to be from 1884 to 1920, wrote in 1962 that during that period, at least in Hodgeman County, posts sold for about thirty to thirty-five cents apiece, including hauling them distances of one to four miles. Margaret Haun Raser, however, thought that during the early period the price in Hodgeman County was somewhat less.

Prices the State Board of Agriculture quoted for posts in 1879 and 1880 (*Second Biennial Report*) probably were for wood posts. The prices differed little from some of the quotations, given by various sources, for stone posts a few years later. If similar prices for wood posts prevailed for the next two or three decades (subsequent biennial reports do not quote prices for posts), then cost was not a major reason for using stone instead of wood posts. On the other hand, fencing in north-central Kansas was in its infancy in 1879 and 1880, so probably the timber supply along streams still met demands for posts, and optimism about growing hedge fences prevailed. A few years later, when more attention was given to fencing and preferences turned to posts and wire, the scarcity of native timber undoubtedly made its impact, and the price of shipped-in post timber could have been "out of sight," as one old-timer averred. That, plus the stone's durability and ability to withstand prairie fires, put stone posts in the reigning position.

Stone posts reigned, that is, until about 1920, when home-grown hedge posts could meet demands in some areas; shipped-in, treated wood posts had become available at prices that suited the farmer's budget; and the farmer had less time (quarrying rock was a long, slow process). Then, stone-post quarrying began to decline, and the price rose. Arthur Jepsen, among those who continued to quarry posts, quoted fifty cents a post in the 1920s.

In 1926 the county engineer of Ellis County, as reported in a Kansas Geological Survey bulletin, *Geologic Investigations in Western Kansas* by N. W. Bass, estimated that stone posts at that time were worth fifty cents each. By 1932, according to another Kansas Geological Survey bulletin, *The Geology of Ness and Hodgeman Counties, Kansas* by Rycroft G. Moss, "stone posts now cost one dollar each, so most new posts are of wood and have been shipped in." The report noted, however, that more than three thousand miles of stone posts then were in use in the two counties and that their combined worth was "about three-quarters of a million dollars."

If other counties in post-rock land at that time had a comparable display of stone posts, and indications are that they did, anyone with the patience and time probably could have counted more than five million posts, worth as many dollars, in north-central Kansas.

STONE POSTS BY THE MILE

After a good start in the 1880s, post quarrying spread to almost every part of the region where the brown-streaked source rock could be found. Surely before 1900 even the casual observer could map the area simply by moving from fence line to fence line or post quarry to post quarry, except perhaps in the northeasternmost part. Throughout the land, on either side of stream or gully, nearly every hillside had its creamy-buff scars that marked quarry sites, commonly accented with piles of split-out posts ready to be hauled to the fence lines. And, as Erasmus Haworth observed *(Mineral Resources of Kansas for 1897)*, "Travelers passing from east to west along almost any railroad line in the state can notice large fields and pastures fenced entirely by fastening the wire fencing to those stone posts."

When F. W. Cragin, geologist with the United States Geological Survey, did some geologic mapping in the area in the 1890s, he could not escape the sight of mile after mile of stone posts in mile after mile of fence lines. It was just as obvious to him which rock ledge was being used so extensively for the posts. In fact the bed was so distinct he could use it as a "marker bed" to help in his mapping. Although he thought of calling it Downs limestone, probably in honor of Downs at the eastern edge of Osborne County, he also referred to it in *Stratigraphy of the Platte Series or The Upper Cretaceous of the Plains,* published in 1896, as "Fencepost limestone." Fencepost limestone was the name that stuck, appearing thereafter in various geologic accounts until it became a part of the accepted nomenclature. Simultaneously, the designation "post rock" was gaining popularity, and now that informal name is also universally accepted. Post-rock posts are generally uniformly shaped, have graceful lines, are sturdy and colorful.

Lest we here seem to imply that only the post-rock bed was used for fence posts, we hasten to add that a few other limestone beds of the Cretaceous-age Greenhorn limestone unit (Chapter 5) were being used to some extent as posts. One, which for obvious reasons was dubbed the "Shellrock" bed, has been used especially in Cloud County. The grayish-white "Shellrock" posts are not so colorful or shapely as those of Fencepost limestone. Neither do they withstand weathering so effectively. Occasionally a few other ledges of the Greenhorn unit have been tapped for fenceposts. And along a few roadways in Lincoln County, we observed a few posts that were shaped from brown Dakota sandstone and made colorful by the yellow-green lichens growing on them.

In the western fringe zone of the post-rock area, especially in northwestern Ellis County and adjacent Rooks County, we observed fence posts chiseled from the massive, yellow Fort Hays chalk, but they are neither so graceful nor so serviceable as the post-rock posts. Most of the Fort Hays chalk posts may be up to twice as wide as the post-rock posts and are less tailored. They tend to crumble or break under prolonged attack by weathering, even in this dry climate.

By the end of World War I, stone posts were taken for granted in north-central Kansas, but at the same time the national economy was changing and regional isolation fading. Post-rock country's farmers, equipped with better machinery, were farming more efficiently, and the area's surviving towns were welcoming new trade opportunities. There seemed never enough time to do whatever seemed important to do. Thus, though the stone posts were there—around pasture and field—doing their job and accepted as belonging there, few residents were willing, or could afford, to take the time to continue to quarry stone posts. Besides, as better transportation brought the outsider closer, other types of posts could be obtained with relative ease and at a price that was not formidable. Thomas Frusher explained: "After 1920 the usefulness of the limestone fence post began to decline as the treated timber came into use.... It lasted a long time, was cheaper and considerably easier to handle. Also getting them [the stone posts] out of the quarries was a long slow process."

Myron Chapman of Beloit pointed out another factor: In some parts of the region there was a tendency toward growing more crops and keeping

Posts of Shellrock come in various shapes and sizes; this one, west of Concordia, Cloud County, also exhibits an odd wire attachment (upper left). Some tall posts of Fencepost limestone on the "McKinney" place in northwestern Lincoln County (upper right). They are the remains of a barn complex destroyed years ago by lightning. Note the centrally placed brown streaks in these posts.

Massive posts of Fort Hays chalk, northwestern Ellis County.

no cattle (or keeping them in feedlots), thus lessening the need for fencing. He added that with no fences for machinery to get caught in, farming was easier.

Despite all that, stone-post quarrying and setting did not halt in the 1920s or even in the 1930s, and annually at least up to World War II a few more miles of fence lines with stone posts were added in a few localities. Various sources, including geological reports on post-rock counties issued since the 1920s, refer to the "thousands of stone posts still in use." As recently as 1950, according to one estimate, thirty to forty thousand miles of them could be traced throughout north-central Kansas.

Undoubtedly, many fewer stone posts are intact now than in 1950, and the miles of fence lines certainly have been reduced. For example, along many widened roadways where post-rock fence lines were removed to permit the widening, we now see either no fence lines or new fence lines with wood or steel posts. In some places, however, the stone posts were reset in the new fence lines.

Still, in many places stone posts continue to march along mile after mile, as if determined to defend the region's title, "Land of the Post Rock."

That is the saga of the stone posts.

Stone posts "marching along" near the west end of Wilson Reservoir dam, Russell County.

4
POST ROCK'S USES
IN PERSPECTIVE

Obviously, by name and use, Fencepost limestone has an affinity with fence posts. The relation borders on the unique, excuse enough to concentrate on the product, stone posts, to the neglect of the limestone's other uses. Yet the limestone's use for posts, however significant, has complemented, rather than dominated, its prior and primary use as building stone. Additionally, the rock has served north-central Kansas residents in many other ways. Thus, post rock has a broad perspective.

Even the settlers with little knowledge, initially, of how to quarry or to lay stone were inclined to use loose stones or, if they had a tool to use as a sledge, to break off blocks from ledges to line dug wells and cellars, to form inside walls or outside fronts of dugouts, to construct fireplaces, to make steps and small porches and stepping-stone walks and the like. The Rev. Raphael Engel of Hays in recent correspondence on early uses of stone mentioned that there were "hundreds of dug wells fifty and more feet deep walled out with rock from bottom to top" and that "nearly every homestead had a vaulted cellar, also used as tornado shelter." The diary of an early settler, J. Z. Springer, who later became postmaster of Lincoln, contains one of the first recorded references to the use of native rock in the Lincoln vicinity. As printed in the *Lincoln Sentinel* in November 1933, the sole entry for June 13, 1872, was: "Hauled stone for a pig pen."

In the 1870s and 1880s virtually every community in north-central Kansas included among its residents stoneworkers from the "old country" or "back east." That assured no lack of available knowledge about the building potential of post rock and certain other chalky limestone ledges that together were about the most prominent landmarks of the region. Cooperative work on community projects that involved the use of stone gave those with little or no experience in masonry construction a chance to observe the trained craftsmen and to develop some masonry skills. Margaret Evans Caldwell of Hanston said that George Bradshaw, a member of a Negro exodus group that settled in Hodgeman County after the Civil War, learned the masonry trade by helping stonemasons with stonework on the Mudge Ranch. He and his sons thereafter built a number of stone houses and barns in Hodgeman County.

BUILDING STONE AND ITS COMPETITORS

As soon as they could find the time, some homesteaders began to build their "dream homes" of this elegant limestone. Lumber also was used, to be sure, but as one early-day resident explained, "It took money to build a frame house but only labor to build one of stone."

It took only labor, that is, for the settler skilled enough to lay the stone himself. But then, many were prompted to acquire some skill, for few could afford to hire a trained stonemason, a craftsman entitled to top wages—$1.50

House of Fencepost limestone built in 1870 by Thure Wohlfort on his homestead south of Scandia, Republic County. Simple in architectural style, it stands in the shadow of a larger stone house, Victorian in style, built by Wohlfort in 1893.

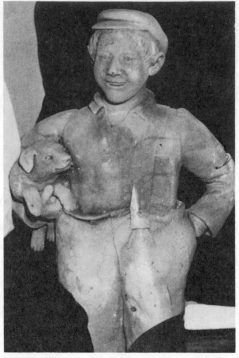

Grave marker of Fencepost limestone in the Bunker Hill cemetery.

Life-size statue of boy with dog carved of limestone by Joseph Hill. *Photograph courtesy of Harold Dwyer.*

to $3.00 a day in 1880 (as determined from quotations in county descriptions in the *Second Biennial Report* of the State Board of Agriculture). By the mid 1880s, improvements on claims in every part of the Smoky Hills Region included not only stone houses but also stone farm outbuildings. Commonly other stonework was in evidence: stone foundations and footings, stone well curbs, stone walls, and hollowed-out slabs used as drinking vessels and feeding troughs for farm animals, not to mention fence posts, gate posts, hitching posts, clothesline posts, and a few telephone poles. Some stonework, including ornamental pieces and tombstones, revealed the carving ability of some of the early craftsmen. A few, among them Joseph Hill, who farmed in the Beloit vicinity, found time for a hobby of sculpturing, using post rock as the material. A boy holding a dog was one of Hill's life-size statues.

In the emerging towns, native rock was being used extensively as a building material, and business districts were made inviting to customers by flagstone sidewalks and stone hitching posts along the entranceways to the stone stores. Dealing in stone evolved into a major business. Some of the products achieved more than local acclaim, as revealed in Mrs. Seymour Mooney's letter of November 18, 1888, to the *Plattsburgh* (New York) *Republican,* in which she described a prize-winning exhibit of the Bazine Stone

Hays in 1884. Note stone posts in front of store of stone. Part of the sidewalk is flag-stone; a stone Presbyterian church built in 1879 dominates the background. *From an old print donated to the Ellis County Historical Society by the J. B. Basgall family.*

and Lime Company at the Kansas City Exposition in 1888 and mentioned especially "building stone, stone centre tables, grave stones, hitching posts, fence posts and rough stone as taken from the quarry." Local newspaper editors were wont to speak with exaggerated pride of their limestone-built towns. For example, we found in the *Beloit Weekly Record,* November 23, 1877: "The reader should come to Beloit, the capitol [sic] of the county and see a model young city built of this matchless material [limestone]. Here are churches, hotels, stores, shops, school houses, bank, residences and barns of stone, from the elegant flagging, three inches thick to noble dimension blocks. It takes the finest finish, even to a brilliant polish. . . . They have built up Beloit with unusual elegance and solidity."

Obviously a key resource in the area's economy, post rock in use just as obviously was beginning to symbolize progress. All of which lends some credence to a statement made by one early-day settler, "In those days every man was a stonemason."

Old newspaper accounts and early Kansas Board of Agriculture reports also mention the use of "native lime" for mortar and plaster in masonry construction. The *Fourth Biennial Report* (1881–82) states for Ellis County: "Native lime used by plasterers and masons. When not exposed to water, equal to quick lime. Used in making mortar and plaster. Used for inside walls." And for Osborne County: "Native lime substance found in many places . . . used extensively in building. . . . Considered to be equal to quick lime when sheltered from moisture." The *Fifth Biennial Report* (1883–84) contains this for Hodgeman County: "Native lime used largely in building; found with magnesian limestone."[1]

It would seem almost providential that the mortar needed to lay building blocks would occur, as it did, in conjunction with the building material. Some of this "native lime," occurring as limestone dust or shaly lime above or below the limestone ledge, was used dry—unslaked (without burning or adding water). When placed between the blocks in the walls, it would draw moisture, thus sealing the joints. Much of the lime, however, was obtained by burning broken pieces of limestone in crude kilns along creek banks. Burning lime mortar and plaster in fact was one of the first industries to evolve in association with limestone quarrying and the building trade. In Russell County, the mortar sold for fifty cents a bushel. The centennial edition of the *Russell Daily News,* June 1, 1971, states:

> The Johnnie Schmitt farm near Gorham still offers evidence of one of the earliest Russell County industries. The remains of two old limestone kilns can still be seen on the banks of Big Creek.
>
> Holes were dug into the bank six feet deep and about the same circumference. Shell rocks of limestone were piled into the opening and fired with the wood and brush that grew along the bank until the powered lime was formed.
>
> This lime was used for plaster, being mixed with animal hair, and for other building jobs around Gorham.

Schmitt told us that one of the kilns, "the best that I have seen," could be cleaned out easily and preserved. Another article in that centennial edition tells how the Gottlieb Schwartz family, among the first settlers in the Gorham vicinity, finished the interior walls of their dugout in a creek bank: "Mrs. Schwartz and the children gathered dried buffalo grass, mixed it with native clay gumbo lime and water to make plaster for the room. She had learned this art in her native Austria. The plastering, with occasional whitewashing, lasted the seven years the Schwartzes lived in the dugout." Gilbert Schwartz, a grandson now living in Gorham, said that Mr. and Mrs. Gottlieb Schwartz came to the area in 1878.

Eight miles south of Russell and about eight miles southeast of the Schmitt farm, R. B. Landon took a homestead in 1871. Almost immediately he began picking up stones from the hillsides for a house and simultaneously began burning lime for mortar. When completed, the Landon house was the only one on the road between Russell and Great Bend, according to Melvin Flegler, who lives on the property, now owned by the Andrew Flegler heirs. Literally built on the creek bank, the house measured twenty by thirty feet and had six windows and two doors; a story below the main level extended into the bank slope. The stone house, along with a stone wall and stone steps on the creek-bank side, is still standing.

The Mehls noted that within their memory lime kilns could be found along stream banks in many places in Mitchell County. Jane Mehl wrote in a letter:

> According to information we gathered from Dad, in this area the kilns were dug approximately 20 feet deep, 6–8 feet in diameter. They were lined all around (like a well) with limestone rock, except for a hole at the bottom large enough to keep a hot fire going. The constant heat (probably kept burning for a week or more) baked the limestone. The ashes from the fire were cleaned out of the kiln; then, starting at the top of the walled up rock, the rocks were knocked down into the bottom of the kiln and as they hit the bottom they pulverized.

Mrs. Inez Ernzen, Mark Torr, and Arvilla (Mrs. Paul W.) Heiman, Mitchell County residents interested in archeology, have succeeded in adding to the list of historic archeological sites in Kansas a lime kiln on Mill Creek south of Cawker City. The kiln, eleven feet deep and twelve feet across at the bottom, is the state's largest kiln that has been reported up to 1972, according to Dr. Thomas Witty, state archeologist.

Lime kiln as sketched by David Mehl, teen-age son of Clarence B. Mehl, in 1973.

While native rock emerged naturally as a building resource, other materials had their day and their place in the region's early development. Along some of the streams logs were felled for building; in some communities clay was fired into building brick; and when the railroads began shipping in lumber, some settlers with money in their pockets chose to buy and use lumber. Many communities for expediency erected frame business and public buildings. In addition, in most building projects some wood was required.

For an example, let us consider the building materials of the schoolhouses existing in 1875 and in 1880 in five representative post-rock counties: Republic, Cloud, Mitchell, Lincoln, Russell. The accompanying table shows that, overall, frame construction predominated but that stone was a major contender, more so in 1875 than in 1880. Only a few schoolhouses were of brick, and those of log must have been temporary in that there were more of them in 1875 than in 1880.

In analyzing the data in the table, several factors should be kept in mind. Of the counties listed, Republic and Cloud, the easternmost, were

NUMBER OF SCHOOLHOUSES IN FIVE POST-ROCK COUNTIES, BY BUILDING MATERIAL, 1875 AND 1880

	Stone	Frame	Log	Brick	All school-houses
Republic County (1868):					
1875	16	50	11	2	79
1880	18	77	3	2	100
Cloud County (1866):					
1875	11	26	4	0	41
1880	27	65	0	1	93
Mitchell County (1870):					
1875	21	12	16	1	50
1880	31	55	3	0	89
Lincoln County (1870):					
1875	4	10	3	0	17
1880	17	29	4	0	50
Russell County (1872):					
1875	8	0	0	0	8
1880	17	10	8	0	35

Dates that counties were organized given in parentheses.

Source: State Board of Agriculture: *Fourth Annual Report*, 1875; *Second Biennial Report*, 1879–1880.

settled first and probably had the most native woodland. Russell, the westernmost, was settled last and had the least timber. Also, in Russell County, as pointed out by Mrs. Blanche Lange of Bunker Hill, the Union Pacific Railroad, which crossed the county in 1867, stripped the timber along the streams for fuel to run the locomotives. (Russell was the only county of the five listed that had railway service before the late 1870s.)

The figures in the table indicate that except for Lincoln County the proportion of wood to stone schoolhouses increased between 1875 and 1880. By 1880 the shift was clearly toward lumber. A reasonable explanation for that trend might be that at first stone, which was available where lumber was scarce, could be used to erect a schoolhouse at almost no cost, provided the settlers contributed the labor. They seemed willing to do that before they became deeply involved proving up their homesteads, and most wanted their children to have access to schools as soon as possible. Later, when making a living on claims was a full-time struggle, there were more settlers to share the cost of building materials; at the same time railroad towns with lumber yards were within hauling distance. Because it was faster to erect frame than stone buildings, settlers were willing to contribute toward the purchase of lumber. Even so, if we interpret our reading of historical accounts correctly, many of those frame structures were considered to be temporary.

A place to worship, as well as a place to learn to read and write, was important to the pioneers. Consequently, churches (both frame and stone) were among the first community buildings. Many congregations, however, seemed willing to meet in homes or other temporary quarters, such as schoolhouses, while taking several years to plan and erect suitable houses of worship, preferably of stone.

Whatever trend there might have been toward building with lumber, stone still continued to be in demand as a material of permanence. The risk of fire, one of the most dreaded hazards of the plains, increased with the coming of the very railroads that hastened the country's development by bringing lumber to erect business buildings and other structures quickly. Sparks from the wood-burning or coal-burning locomotives moving across the prairies could start prairie fires that might wipe out business blocks of frame buildings. Thus the railroads were in part responsible for a refocus on native stone as a durable, fire-resistant building material.

The February 28, 1884, issue of the *Russell Record,* in a question-and-answer article directed to "the large and increasing number of Eastern readers . . . making . . . inquiries concerning our natural advantages with a view to settling among us," listed only stone as a building material, pointing out "resistance to fire" as one of its advantages:

> There are two varieties of building stone in abundance, cropping out
> along the bluffs near the top, and very easily quarried. One kind is a

soft yellow Magnesian Limestone, with no grit in it whatever, which is easily worked with hand saws and carpenter's planes into any required shape, and hardens when exposed to the weather. A wall of this stone stood the test of the worst fire Russell ever had without any serious damage. The other principal variety is a grayish stone of medium hardness, which is much used for building, and also makes excellent walks.

In a supplement to the *Public Record,* Cawker City, 1885, an item tells of Cawker City's move from wood to stone business buildings:

A. W. Smith came to Cawker City in September, 1871, and began the bakery business on the south side of Wisconsin Avenue. A fire occurred April 1, '79 destroying five buildings, including Mr. Smith's, but he immediately put up a two story stone building on the same lot. . . .

The fire above mentioned, though a loss to individuals, was the means of causing the creation of more substantial buildings and the passage of a fire ordinance forbidding any but stone buildings being erected within prescribed limits, and Cawker now boasts of having the most substantial business houses in the northwest.

POST ROCK: BUILDERS' CHOICE, THE LATE 1800s AND EARLY 1900s

Because it had demonstrated its ability to resist fire and other destructive forces, was elegant, and had other desirable characteristics as well, Fencepost limestone began increasing in stature as a building material. That much of north-central Kansas had entered a boom period by the mid 1880s helped. The hardy pioneers who had stayed on—toughened by a series of setbacks brought on by such disasters as droughts, blizzards, grasshopper invasions, and prairie fires—were ready to build for the future. As they built, the regional economy for the next three or four decades was linked closely to the quarrying of limestone, primarily as building stone and secondarily for fence posts and other uses.

The editors of the *Lincoln Beacon* (March 3, 1887), in a somewhat poetic article on the grandeur of Lincoln County, subtly linked that county's future with the quarrying of post rock: "One can even ride or walk over the very edge of the bluffs into the valleys, or vice versa, as we have, thousands of times, never suspecting that a few weeks, months, or years more [and] a quarry of beautiful creamy white limestone will be opened directly under the footprints he left." Before the turn of the century, many a quarry of that "creamy white limestone" had appeared where footprints had been and had not been along the bluffs, or escarpments, in Lincoln County and throughout post-rock land. In fact, from the 1870s into the second decade of the 1900s, each year new quarries were opened on both sides of the major streams and their tributary valleys from the Republican River in Republic County to Buckner Creek in Hodgeman County.

LAND OF THE POST ROCK

Quarries in the vicinities of Dorrance and Bunker Hill, along the Kansas (Union) Pacific Railroad, were among the first to provide post rock for more than local use. This notation on Russell County in the *First Biennial Report* (1877–78) of the State Board of Agriculture undoubtedly referred to this area: "Abundance of excellent building stone, 6 to 9 inches in thickness, easy of access. Has been shipped east and west on the Kansas Pacific for building purposes." Floyd Duncan said that in the area where he lived, about midway between Lucas and Bunker Hill, there were at least a dozen quarries in the early 1900s. Concomitantly, during that period in this area, today dominated by the Wilson Reservoir, several luxurious ranch

"Creamy-white limestone along bluff" in vicinity of Sylvan Grove, Lincoln County, and a lake in an old stone quarry in the vicinity of Schoenchen, Ellis County.

A. G. T. Cooper ranch, in Wilson Reservoir area eight miles north of Dorrance, Russell County, as photographed by Leslie Halbe in 1911. *From the Halbe Collection, Kansas State Historical Society.*

houses of post rock were erected. One of the most elaborate was that of A. G. T. Cooper, described in the *Kansas City Star,* October 22, 1911, as "the largest and most expensive farmhouse [it cost $20,000 plus $3,000 for furnishings] in western Kansas."

For several decades quarries were conspicuous along the bluffs of Big Timber Creek east of Liebenthal, along the Smoky Hill River in the vicinities of Schoenchen and Pfeifer, and along Big Creek south of Victoria.

Limestone quarrying along the escarpments bordering Walnut Creek and tributaries near Bazine provided that community with one of its major businesses from the late 1870s to about 1920. On February 13, 1889, Mrs. Seymour Mooney of Bazine wrote to the *Plattsburgh* (New York) *Republican*: "The limestone company of this place [The Bazine Stone and Lime Company] are shipping car loads of stone every week to various points which gives employment to a goodly number of men, who are ready and willing to work for a dollar and a quarter a day and board themselves." On January 29, 1901, she wrote to that New York paper:

> One of the great revenues of Ness County comes from the quarries; the ledges of rock are inexhaustible, and the product is always in demand. But few of the towns in the county have so fruitful a mine as Bazine hence the filling of cars at the stations goes on the year round with posts for fencing, blocks for paving and stone for all building purposes are shipped to various points, perhaps hundreds of miles away. When a man gets short of money he takes his team and with it makes two dollars per day at the Quarries.

In her 1901 item, Mrs. Mooney revealed not only that Bazine stone was marketed widely but also that the quarry products were diverse. It also appears that wages were rising. Here, as well as elsewhere in post-rock country, stone posts almost always were quarried along with the major product, building block. Commonly the quarries also produced flagstone for walks and a variety of other byproducts, including gravestones, hitching posts, paving blocks, and stone for ornamental pieces.

Emil Mauth told us that during the quarrying heyday, ten or more quarries were operating in the vicinity of Bazine. As a stonemason working in some of those quarries between 1900 and 1930, he prepared a lot of building block, using a chisel and stone hammer to shape blocks into a standard two feet by eight inches by eight inches. Most blocks to be used in walls he made pitched-face, as "that type most commonly was used during this era." He used a "tooth chisel" to dress sills and headers (stone over doors and windows) for buildings. At first he used only hand tools; later, a rock-dressing machine for some of the work. The finished building block sold for five to eight cents a cubic foot.

Elsewhere in post-rock country, quarries on the surrounding bluffs early became a part of the view from developing towns, including all those along the Solomon Valley in Mitchell County. The *Beloit Courier* of May 1, 1879, mentioned that limestone "is to be found in immense quarries, within from one to three miles of the city. . . . It enters largely into the construction of business houses here. . . . It costs about $10 per cord in the wall."[2]

About 1885, when the building boom was on in Jetmore, a number of limestone quarries were opened on the nearby slopes along Buckner Creek and for several years were a source of many loads of stone for Jetmore buildings, where business is still being transacted today.

Clarence Mehl noted that about 1900, "building block was priced by the wagon load, and a wagon load did not mean it was piled high in the wagon box, but rather in two layers in the bottom of the box. A 'load' weighed about a ton, and the price was approximately $1.50 per ton. That was 'rock in the raw' so to speak. It was hauled to the building site and there dressed by a stonemason."

An example of a family quarrying business near the turn of the century was an operation about seven miles northwest of the German settlement of Holyrood and across the Ellsworth County border into Barton County. There, on the divide between the Smoky Hill and Arkansas rivers, John Altenbaumer, a blacksmith, opened two quarries (one for building stone primarily and the other principally for fence posts) on the 400-acre farm he and his wife Christena bought in 1897 from a railroad company for $7.00 an acre. Mrs. Matilda Adams, a daughter living in Lyons, said that two brothers "on a visit from the East" helped in the business and received about $12.00 a month plus room, board, and laundry. The Altenbaumers delivered both

Jetmore building erected in the 1880s (and dismantled in 1972) and quarry site south of town where much of the rock for Jetmore's business buildings was obtained.

building block and stone posts throughout Barton, Ellsworth, and Rice counties, receiving twenty cents and up for building block (each weighing fifty to sixty pounds) and about fifty cents for the posts; prices depended on size of blocks or posts and distance of delivery. John Altenbaumer used some of the stone he quarried to build a barn and the basement and foundation of a house on the homestead. The Altenbaumers sold the farm (with its quarries) for $17.00 an acre in 1902; it is now owned and lived on by Mrs. Mary Zajic and her son Derald.

Southern Barton County, eastern Ellsworth County, and almost all of Rice County are south and east of post rock's outcrop area. Thus, the Altenbaumers delivered their post-rock products thirty to forty miles beyond the boundary of the region proper. Similarly, post rock was delivered (by others) to the northwest, as evidenced by post-rock buildings and fence posts in northwestern Ellis, southern and eastern Rooks, northwestern Osborne, southern and eastern Smith, and northwestern Jewell counties. We observed, for example, a jail of post rock on the courthouse square in Smith Center, a post-rock library (formerly a school) in Burr Oak, and the St. Joseph Catholic Church in Damar. Those three towns are well beyond the Fencepost outcrop area. Though Osborne, like Hays, is close enough to be considered a part of the Land of the Post Rock, much of the post rock used in Osborne's buildings, for example the courthouse, was quarried in counties to the south and east of Osborne County. In this area adjacent to post-rock country on the northwest, Fort Hays chalk also has been used extensively for building—and to some extent for fence posts. Fort Hays stone generally can be distinguished from Fencepost by its uniform color, its softness and tendency to crumble when it weathers, and its massiveness— large building blocks and huge posts.

To obtain building block, fence posts, or other products, the quarrying process was the same: Holes were drilled about four or five inches deep into the rock and nine to twelve inches apart along a line marked for splitting; feathers and wedges were placed in the holes; and tapping the wedges lightly with a stone hammer split out the slabs, posts, or blocks. By World War I, some quarrymen were using gasoline engines for power to drive the drill bit into the rock, derricks and hoisting devices to load the rock, and various home-made mechanical gadgets to make their work easier.[3]

It is probably correct to say that until well into the first decade of the twentieth century most of the native limestone used in the region's buildings was quarry stone—blocks used as delivered from the quarry without further trim. Before World War I, however, many blocks or slabs destined for building stone were cut into specified lengths and finished, with hand tools, at the building site. Blocks might be smoothed on all faces or they might be dressed or made "rough" on the side to be exposed in the building wall. Contacts between blocks laid in the wall commonly were irregular; in the building trade, such blocks were called "rough hewn," or simply "rough."

The rough-hewn blocks of a few old buildings show no drill marks that would indicate the use of feathers and wedges in the quarrying process. A good guess is that those blocks were sledged out. A notable example of such a building is the Bunkerhill Museum, constructed in 1880 as a Lutheran church.

East of Beloit in the Scottsville vicinity in northeastern Mitchell County and the adjacent corner areas of Jewell, Republic, and Cloud counties, the walls of many buildings are of sledged-out blocks that had been split along the rock layer's brown streak, forming two blocks. In Scottsville we examined an old community building that shows split blocks on two walls, unsplit blocks on the other two. We had learned from Bird Abram that his

Stone quarries in the Bunker Hill and Dorrance vicinities, as photographed by Leslie Halbe in 1911. *From the Halbe Collection, Kansas State Historical Society.*

Stone quarry on the Martin Jensen farm, northwestern Lincoln County, in 1913 or 1914. Note hoist used to lift stone on wagon bed. *Photograph courtesy of Mr. and Mrs. Arthur Jepsen.*

August Ruppenthal ranch, on the south side of the Saline River north of Dorrance, as Leslie Halbe saw it in 1911. The walls of the barn show quarry stone with a pitched face; of the house, rough-hewn stone with a relatively smooth face. *From the Halbe Collection, Kansas State Historical Society.*

Fencepost bed along the Solomon River near Beloit, showing its tendency to split along the brown streak; and a section of a side wall and back wall of the community building in Scottsville, showing the use of unsplit and split post rock.

father, C. W. Abram, and Samuel Carter had sledged out split blocks in that area, but we were not sure why until we examined the rock in its outcrop in several places along the Solomon east of Beloit. We noted that the Fence-post layer there has a natural tendency to split along the center of the brown streak, apparently the reason for the split blocks. Some blocks resembled brick and some "flagging."

To make flagging from a slab of post rock, the quarryman split it along the brown streak using a wedge lightly tapped by a hammer. Used exten-sively as sidewalks, with either the brown streak or the lighter-buff face showing, flagstones were a byproduct of quarrying for building stone and fence posts. Golden Morris said that slabs too fractured for either building blocks or posts were set aside for flagging. The flagstones, which were at least as wide as a normal sidewalk, were hauled to town, he said, by looping chains around them and hanging them to the running gear of a wagon.

Flagstone walk in Lincoln.

In some localities Greenhorn limestone layers thinner than the Fence-post bed, including one or more referred to in the geological literature as the "Flagstone bed," were used for sidewalks. But whether split Fencepost slabs or thin slabs of another limestone bed were used, flagstone sidewalks were the pride of many post-rock towns. A promotional publication, *His-torical and Descriptive Review of Kansas,* issued in 1890, praises both Be-loit's stone business buildings and, with enthusiasm, the town's sidewalks: "Beloit takes great pride in her . . . miles of stone sidewalks that will remain firm under the hurrying footsteps of the generations as they succeed each other."

In a centennial booklet, *History of Ellsworth County,* John F. Choitz comments that Ellsworth once boasted of having in front of its Grand Cen-tral Hotel (built of brick in the early 1870s) the "best sidewalk in Kansas . . . a slab of magnesian limestone twelve feet wide."

LAND OF THE POST ROCK

An item in the *Bunker Hill Banner,* March 17, 1882, stated: "The large stones used in the walk between Hill House and Eyler Store [currently the town's grocery store] were hauled from the local stone quarry. They were fastened with chains to the running gears of the rock wagons."

The *Russell Daily News,* centennial edition, June 1, 1971, notes that at the beginning of the 1890 decade, the sidewalks of Russell's business district were of "large flat rocks quarried nearby."

Fred Wessling, Beloit carpenter who has used a lot of native limestone in building, said, "Flagging stone was used at an early date as floors in barns and cellars." Later, flagstone became popular for patios and for yard landscaping.

Some flagging also has been used as building stone. A good example of the use of a three-inch flagstone from the Greenhorn unit is in a vacant, two-story house on property owned by Leo Wallace about a mile north of Barnard. The flags are so well fitted that the walls are in excellent condition though the house probably was built before 1900, according to Frank Dowlin, who once lived there. An example of the use of split Fencepost limestone as flagging is in a rural schoolhouse, built in 1883, two miles east of Beloit. The deep buff of the flagging in the schoolhouse walls is accentuated by white Shellrock used for trim and by unsplit post rock used as porch slabs

District No. 3 schoolhouse east of Beloit, Mitchell County; built in 1883 of split Fencepost limestone. *Photograph by Alan Houghton.*

and steps. The school closed in the 1960s, but the sturdy building still serves a purpose. Alan Houghton, publisher of the *Beloit Daily Call,* mentioned that the Mitchell County Association for Retarded Children holds meetings there regularly.

In various communities in the region's westernmost counties, including Russell, Ellis, and Ness, limestone ledges exposed high along the bluffs have been the source of "stone brick" for residences and other buildings. The limestone is quite similar in appearance to the Fencepost but is softer, chalkier, and geologically younger. The *Russell County Record* of January 30, 1879, revealed the magnitude of Russell's stone-brick industry based on the use of that chalky limestone: "There are now three mills—two run by horse power, and one by steam, sawing brick from the soft stone of the county, and another will soon be in operation to be run by a 25 foot wind wheel." The *Russell County Record,* March 27, 1879, had this item on Bunker Hill: "Smith & Brown . . . [are] running a brick-making machine, sawing brick from the softer variety of stone which is found so abundantly in Russell County."

Both Bunker Hill and Russell provide examples of houses built of stone brick during this early period. The one on the old Charles Shaffer homestead north of Bunker Hill has one wing built, about 1872, on the quarry-stone foundation of a dugout, according to Ray Shaffer, only surviving member of that family in 1973.

After 1900 building and maintaining public roads became increasingly important to the region, as well as to the rest of the state. Post rock was called upon for yet other services: as a material for bridges, for road embankments to control erosion, and, along with Shellrock, in a crushed form for road surfacing. In building bridges—both for public roads and for railroads—the stone arch emerged as a popular architectural form. One of the first built in post-rock country was across Paradise Creek, near Paradise. Completed in January 1902, it had one large arch, fifty feet wide, and it cost $4,070.45, twice what any other type bridge would have cost then. The cost was justified, said the *Russell Record,* because "no person now living in the county will see the day when it will need repairs, other than repointing." In 1973 this bridge was still being used.

By September of 1902 Russell County could boast of a second stone-arch bridge, built across Big Creek south of Gorham, at a cost of $2,477.86 ($219.54 for filling and earthwork and the rest for masonry). That bridge was larger than the one across Paradise Creek; it had two arches, each 36 feet wide; it was 225 feet long and 18 feet from low water to top; and the height of the arches was 18 feet. The *Russell Record,* September 27, 1902, said of it:

> With the exception of the cement every dollar's worth of material was produced in Russell County. The stone were quarried only a few

Stone-arch bridge across Big Creek south of Gorham, Russell County; built in 1902.

hundred yards from the spot. Some of them are from six to eight feet in length. The foundation is solid, and there is no reason to doubt that it will stand for hundreds of years and cost little or nothing for repairs. With the exception of two skilled workmen, the work was all done by Russell County labor, thus keeping nearly the whole amount paid for the bridge in the county.

During the next two decades a number of stone-arch bridges were built of post rock throughout post-rock country and in adjacent areas. About the same time, the rock was being used extensively for stone-arch caves, in demand by area farmers for shelter from storms and tornadoes and for a storage place for farm products. Mrs. Inez Ernzen, schoolteacher in Beloit, said that her father quarried post rock and built both stone-arch bridges and stone-arch caves in Mitchell County between 1905 and 1912. Three of those bridges on West Asher Creek, eastern Mitchell County, are in use today. Rex Hodler, who lives about three miles east of Beloit, probably includes them in this evaluation:

> There are several stone arch culverts or small bridges near here. These are made of dressed post rock. The workmanship is good enough so they are well able to carry the tremendous loads now using our roads (20–30 tons is not uncommon now, whereas the team and wagon were scarcely one-tenth as heavy). They also were some of the few bridges that were capable of carrying the massive weight of the first large steam threshing engines. . . . They will stand another 100 years or more unless man-made machines deliberately destroy them.

Cave on John J. McCurdy estate, with entrance on Raymond homestead, Lincoln County. The roof over the entrance is a solid slab of limestone. *Photographed by C. C. Abercrombie, 1956.*

Floyd Duncan, who helped build caves in the area, explained the method used:

> Stone blocks for the bases of the cave walls were laid to a height of about a foot. Wood forms for the arch (made of 2 x 6s) were set on those wall bases and boards placed over the forms made a solid arch. Stone blocks then simply were laid over those boards. When the laying was completed, the worker knocked out the forms. The stone arch—the cave wall—would stay. Mortar sometimes was put between joints, but it was not necessary, as pressure from the stone would hold the arch in shape.

Duncan explained that for some caves and bridges, the stone blocks were cut into specified wedge shapes so that in the completed arch the blocks would fit precisely.

THE END OF AN ERA AND POST ROCK'S COMEBACK IN THE 1930s

To build a house of permanence with post rock was the dream of many a pioneer couple who began life on the central Kansas prairies in a dugout or soddie. By World War I, a count of the stone dwellings on farms in that region would be proof enough that for many the dream had come true. By 1920, however, building with stone had passed its climax—and so had regional development. Among factors contributing to the decline were the availability of cheaper, easier-to-use building materials, and perhaps the adequacy of individual and public buildings already present. Additionally, many of the old stonemasons were leaving the scene and young men, back from World War I, were qualified for other trades that took them off the homestead and outside the area. The pace of living was geared to the "roaring twenties," and a new type of economy was evolving. Power machinery was beginning to appear on farms; the automobile was giving area residents mobility and the area itself accessibility. The region no longer was an isolated entity, and homesteads were losing their self-sufficiency status.

During the 1920s the rural farming pattern was shifting to fewer farm-steads, though more land was being cultivated. For the sons and daughters of the second and third generations who had chosen to stay on "the old home places," the sod had been broken, the fences were up, and the houses were there to be occupied. (When the drought of the thirties came and the topsoil began to blow, there was little doubt that the sod had been broken.)

Ironically, as "finis" was being written to the great post-rock era, the depression of the 1930s made possible post rock's brief comeback as a major building material. A resource that again came to the aid of regional econ-omy, it was much used in public construction projects funded wholly or partially by the federal Works Projects Administration. Few if any stone posts were quarried and set under the WPA program, but the building proj-ects, with post rock the building material, were many: schools, libraries, city halls and community buildings, bridges, park shelters and recreational facilities, and two courthouses.

Quarrying procedures were basically the same as they had been during the previous half century, but there were a few innovations and improve-ments. Tractors commonly were used to pull the fresno scrapers or "tumble bugs" over the quarry sites to uncover the quarrying ledge. A tumble bug, as described by Wayne Barnett, "was made with a rotary bowl . . . had a trip in 'out' position and one in 'in' position. Trip it once and it would roll half way around to gather up the dirt. When you got ready to dump the dirt, you tripped the latch and it would roll over one time, catch again, and spread the dirt." Gasoline engines—in some areas motors from old washing machines—were incorporated into drilling devices. Some of the devices were wheelbarrow affairs. The motor was attached to the wheel end and the drill bit to the handle end, so that after drilling one hole the quarryman had only to lift the handles and push on to the next spot along the line marked off for drilling.

Most rock was shaped at the construction site, using hand tools and to some extent power equipment, including power saws. It was the practice to saw building block on all faces except the one to be exposed in the wall. That face was chiseled into chipped- or pitched-face ashlar (rectangular with sawed, planed, or rough-faced surfaces not "finished" or accurately sized, as are blocks of cut stone). Some blocks, however, were sawed with precision on all six faces to produce cut stone.

The Jewell County Courthouse at Mankato and the Ellis County Court-house at Hays are two major structures added to the post-rock scene during the depression era. Both are modern and functional, with no clock towers or "gingerbread" trim. Both are excellent examples of the use of cut stone.

Wheelbarrow device with gasoline engine once used in the Mehl quarries in Mitchell County. *Photograph by Alan Houghton, about 1960.*

Ellis County Courthouse, constructed of sawed post rock in 1942.

The courthouse at Mankato was completed in 1938 at a cost of $125 thousand, the one at Hays in 1942 at a cost of $311 thousand (as noted in a table on courthouse construction in the August 1958, issue of the *Kansas Government Journal,* page 450).

Of the schools constructed of post rock in the 1930s, perhaps the largest was the Russell high school, now the Ruppenthal Middle School. Completed in 1938, partially with WPA funding, the building is functional in design and its walls show post rock in pitched-face ashlar construction with cut-stone trim. Architecturally, it represents a return to the simple "prairie" or "plains" style, as described in *Kansas, A Guide to the Sunflower State* (Federal Writers' Project).

An example of a library built of post rock under the wing of the WPA is the Barnard Library at La Crosse. Completed in 1937, it is a modest, well-designed structure of pitched-face ashlar construction that easily could pass for a residence. The city halls at Bison and Dorrance are typical of the blocky style of municipal buildings erected with WPA funding. The main-

Two-arch stone bridge across Hell Creek, southwestern Lincoln County, in high water during flood, late spring of 1973. The bridge was built with WPA funding in the 1930s. *Photographed by Lafe Rees of the* Lincoln Sentinel-Republican.

street bandstand at Spearville probably cannot be considered typical. Shelters, outdoor fireplaces, and other facilities in the city parks of La Crosse and Russell illustrate Fencepost limestone's use in WPA-developed parks and recreational areas.

Of the many water towers built of post rock in towns and on farms in the 1930s, one of the tallest and most visible is the cylindrical structure on the rise at the entrance to Paradise from Kansas 18. The tower, erected of pitched-face blocks in 1937, is sixty-five feet high and has a capacity of fifty thousand gallons, according to Charles J. White, city clerk.

Post rock contributed importantly to highway improvements made with WPA labor in the region in the 1930s. It was a favorite resource for stone-arch bridges—some were built with one, some with two or three, and a few with four arches. A four-arched bridge spans Big Creek south of Walker near the site of old Fort Fletcher (forerunner of Fort Hays). Harvey Roush, who helped build some of the bridges in Lincoln County, said that "for a snug fit" stone for the arches generally was cut as wedges: about nine inches by twelve inches at the top and seven inches by twelve inches at the bottom; depth was seven to eleven inches.

POST ROCK UP TO DATE

The end of World War II is an arbitrary date for the beginning of the modern era of post rock in use. Although quarries by then were few, most operating intermittently, they have been the source of considerable building stone for major regional structures, including especially churches and schools. Some significant changes in quarrying and building techniques have contributed to greater operating efficiency. Modern construction equipment has made it possible to remove additional rock beyond the limits where the pioneers, with their primitive equipment, had had to stop. The use of modern quarrying equipment has speeded up production of building blocks, some of them shaped at the quarry site. Most building block produced today is veneer (defined, along with other building-trade terms, in the note section).[4]

Most stone in old post-rock buildings lies on the bedding plane; that is, as it occurs in nature. In some buildings, however, some of the stone appears as "shiners" (perpendicular to the bedding plane). The shiners tend to deteriorate, whereas the stone in the natural position has good resistance to weathering, at least in the relatively dry climate of north-central Kansas. Some modern quarrymen, believing that the brown streak offers greater resistance to weathering than does the remainder of the rock, saw or split post rock along that streak and then dress the stone so that when laid in the building wall the streak's horizontal surface (in the outcrop) is exposed.

Hubert E. Risser, in *Building Stone in Kansas,* describes the modern process of quarrying post rock:

> The "Fencepost" stone . . . is quarried by sawing directly from the formation. . . . a diamond blade, . . . 30 inches in diameter, is driven by a 30-horsepower gasoline engine and is mounted so that it can be raised or lowered as desired. The unit is mounted on a wheeled carriage, which travels on tracks.
>
> In operation, the overburden is stripped to provide a clean working floor, and a support rail is laid at each end of the area to be cut. A track is laid perpendicular to the two support rails, one end of the track resting on each rail. A slot is cut along the full length of track, and the track is then shifted to a new position on the support rails. Each slot is cut to the full depth of the ledge. Proper spacing of the track position gives slabs of the desired ashlar course height and 8 inches deep. The 8-inch slabs are next split down the center to form 4-inch ashlar veneer.

Joe Prickett, Beloit contractor, operated a modern quarry of that type until the late 1960s, except: "In operating my quarry, I used abrasive blades and electric motors on my saw rigs. I designed and made them myself and they required no water to operate. You have to have water to operate a diamond blade, and that is hardly permissible out in the open in the winter time." His operation consisted of two quarries located about three miles south of Simpson astraddle the Mitchell-Cloud county line. He quarried white Shellrock limestone from the Cloud County side and brown-streaked post rock from the Mitchell County side. He pointed out that he "sawed

Partially quarried Fence-post limestone in quarries (Mitchell County side) south of Simpson, 1952. Observe that in the cutting process, the bed has been split through the brown streak, allowing two blocks instead of one to be produced from the bed's thickness.

this single layer in strips the thickness that we wanted for brown ashlar pattern, and pitched it on the top side as it would lie in the ground. It doesn't slack off from that side." Prickett cut the blocks to various sizes, depending on building specifications, doing some of the finishing near the quarry site and some in a shop at Simpson.

Stone from the Simpson quarries has been used in public buildings and residences in Beloit, Concordia, Clyde, Lincoln, and elsewhere in north central Kansas. One of the most appropriate examples to mention here is the Post Rock Motel complex erected in 1956 at the intersection of Kansas 14 and 18 north of Lincoln. Split blocks from the Simpson post-rock quarry were used for wall veneering in the buildings at the motel site. Measuring about twenty-four inches by two and one-half inches by four inches (the depth, about half the thickness of the original Fencepost bed), they were laid so that the yellowish-brown streak is exposed. Blocks for trim were obtained from the Shellrock quarry.

Currently Prickett uses post rock in various forms, mostly for veneering and decorating. His source of the rock is not always the quarry; it is sometimes building stone, including flagstone, that had been used in other building projects. For example, in veneering the home of Rex Hodler east of Beloit in 1961, Prickett used flagstones that Hodler's grandparents had hauled to their homestead near the Solomon River for walks around their house and in the yard. (Before they built their new house, incidentally, the Hodlers lived in a two-story stone house that had been built in 1872 of pieces of post rock broken out of the quarry with a sledge hammer.)

Quarries operating near Wilson and near Dorrance east of Russell until mid century or later supplied post rock for some modern uses in those vicinities and—as in the past—for building projects as far west as Hays.

When the cornerstone for the Russell County Courthouse was set on July 31, 1902, more than ten thousand people attended the dedication ceremony; Thomas E. Dewey of Abilene was the speaker. The structure was modernized in 1948. *Photograph courtesy of the Russell County Historical Society.*

Interestingly, the Russell County Courthouse at Russell exhibits old-style and new-style construction with post rock from the nearby quarries. Built in 1902 and remodeled in 1948, the courthouse shows in its walls pitched-face ashlar blocks lying on the natural bedding plane (old style) in contrast with smooth (cut) stone of modern design. The result is a new architectural profile: modernistic and functional. The courthouse cost $50 thousand; remodeling it cost three times that much.

In southeastern Ellis County, where once were a score of quarries, in recent years a few operate intermittently to provide building stone for specific projects, in particular in the Hays vicinity but as far west as Colby, as far south as Burdett, and curiously enough east and north as far as Russell and Beloit. George J. Klaus, the principal operator since World War II, has furnished rock for several buildings at Fort Hays Kansas State College, including the newest: Gross Field House. Klaus obtained that stone from a quarry about a mile southwest of Schoenchen. Chipped-face ashlar veneer (over concrete walls) has been used predominantly in the new buildings on the campus. And because the blocks were turned "on edge" to expose the

Fencepost ("Benton") limestone quarry southwest of Schoenchen, Ellis County, operated by George J. Klaus of Hays in 1973. The prominent ledge above the Fencepost bed (quarry floor) is in the Fairport chalk.

bedding plane when laid in the walls, fossils and fossil impressions may be seen in the wall exteriors.

In various communities chipped-face or pitched-face ashlar construction has been popular for modern churches of post rock. Examples include the First Methodist Church at Paradise, completed in 1957; First Presbyterian Church in Beloit, expanded and remodeled in 1948; and the Baptist church in Barnard, completed in 1960. Cut stone also is popular. The newer part of the First Methodist Church in Hays, built in 1948, is of sawed post rock. Post rock for modern churches and other structures has been obtained at various places, sometimes from quarries operated in conjunction with the construction projects.

Beloit's business district in 1958 acquired a somewhat unusual exhibit of post rock and post-rock construction when the Guaranty State Bank was completed. The bank, which was added to in 1972, is unique in that the post rock used in its construction has a pinkish rather than a brownish streak. Furthermore, the pink in the building's walls does not appear as a solid color, but gives a mottled effect because the stone was planed and sanded down to the center of the quarried-out slabs parallel to the bedding plane; that is, to the way the rock strata naturally occur in the quarry. Large square and rectangular blocks of pinkish stone, framed by light-colored prefabricated concrete in various designs—together with the buff brick used in part of the construction—give the bank a geometrical look that emphasizes solidity. A series of bronze plaques depicting historical scenes—the buffalo

Guaranty State Bank, Beloit.

Bas-relief in Fencepost limestone; on base of the "Monarch of the Plains" statue in Frontier Historical Park, Hays. Carved by Peter F. Felten, Jr., to reveal a light-buff surface on a dark-buff (brown), recessed background.

and the Indian, a covered wagon and a pioneer mother, modern crops and livestock—further emphasize the development of post-rock country from pioneer to modern times.

The "pink" rock came from a quarry on the Clarence A. Mehl property about fifteen miles south of Beloit. Clarence B. Mehl, who operates the quarry, told us that so far as is known no other location in post-rock country has a post-rock layer like it. Mehl also processes rock with the brown or dark-buff streak, obtained from a quarry in his pasture. He sells building block "for $1.25 a square foot, dressed and ready for a mason to lay." He provides stone, prepared by modern equipment according to contract specifications, not only for major building projects but also for use in foundations, chimneys, interior fireplaces, interior walls and other interior decorating, and a variety of ornamental pieces.

A few monuments and markers of post rock have been added to the post-rock landscape since 1950. In 1954 a marker with a base of Fencepost limestone (from Ness County) supporting a boulder of native opaline sandstone was unveiled on the old George Washington Carver homestead, one and a half miles south of Beeler. The noted Black scientist homesteaded there in 1886, when he was twenty-two years old. Two years later, he mortgaged his homestead and headed for college. The bronze plaque affixed to the marker lauds Carver as "citizen, scientist, benefactor."

In 1964 a sixteen-foot high monument of post rock and Shellrock appeared on Boyer Hill four miles west of Delphos, northwestern Ottawa County. A project of the Ottawa County Historical Society and constructed by Don D. Ballou, long-time Kansas newspaperman, it commemorates Zebulon Pike's expedition through Kansas. An inscription reads: "Capt. Zebulon M. Pike and exploring party passed this way September 22, 1806, en route to the Republican river Pawnee Indian village to proclaim sovereignty of the United States over this land." One side of the monument contains a bas-relief in concrete that shows Pike on his horse.[5]

The limestone for the monument was obtained partly from the Simpson quarries and partly from a nearby outcrop and abandoned farm buildings. Appropriately, along the road leading up the hill capped by the monument, there is an exposure of almost a complete section of the Greenhorn limestone. The Fencepost bed is not exposed; the monument probably rests on it.

On U.S. 281 (east side) three miles south of Interstate 70 and also on Pioneer Road (west side) four miles east of Russell and three miles south of the Interstate 70 Interchange, we observed stone posts with "BOD 1865" inscribed near their tops. We read this inscription in the cement base of the one on Pioneer Road (the base of the marker on U.S. 281 had so weathered the words could not be deciphered):

Smoky Hill Trail
Butterfield Overland
Despatch
Atchison to Denver
Traversed by Gen. Fremont 1844
First Denver Stagecoach 1859
Retraced and mapped by
Howard C. Raynesford, Ellis, Kansas
Marker placed 1963

These and 135 other such markers—one at every crossing of a major north-south highway—appear along the route of the Butterfield Overland Despatch (BOD) from Ellsworth to Colorado, according to Raynesford's son

Kirk and his niece Edythe Raynesford (Mrs. Charles F.) Speck. The trail, once considered the shortest, albeit the most dangerous, route between Denver and Atchison, is marked because Howard C. Raynesford, an Ellis historian, devoted thirty-five years of research and several hundred miles of hiking to trace the fading ruts of the trail. Raynesford completed tracing the route in 1964, three years before his death at age eighty-nine.[6]

"The Monarch of the Plains" in Frontier Historical Park, at the site of Fort Hays on the south edge of Hays, is one of the latest monuments to appear in post-rock country. Peter F. Felten, Jr., Hays sculptor, completed the project in 1967, a hundred years after the founding of the fort at that site. The mammoth sculpture of the buffalo, or bison, of Indiana limestone, stands on a pedestal of sawed blocks of Fencepost limestone. On three sides of the large sloping pedestal are three bas-reliefs, in slabs of the Fencepost, depicting pioneer scenes associated with the "monarch of the plains." The brown streak has been skillfully worked to produce the reliefs in two colors.

5
WHENCE THE FENCE: GEOLOGY OF POST-ROCK COUNTRY

THE LANDSCAPE—ON THE ROCKS

The stone posts are scattered over a large part of the Smoky Hills, a picturesque hilly area in central and north-central Kansas that sometimes is referred to as the Dissected High Plains because the High Plains once extended eastward across that area. To the west lies the great expanse of the High Plains, to the south the Great Bend Region (or Arkansas River Lowland), and to the east the Flint Hills. The Smoky Hills in the area of stone posts show two distinctive landscape patterns. The eastern part is here referred to as the eastern Smoky Hills or Dakota Country. The western part is known as the Blue Hills. Dakota Country seems haphazardly hummocky, with isolated, irregular hills and mounds rising to various elevations above the valleys. Many of the hills are capped by brown sandstone, but generally the sandstone beds cannot be traced or correlated from one hill to another.

The border between the eastern Smoky Hills and the Flint Hills to the east is vague and cannot be tied down precisely—there is a "no-person's" land between the Flint Hills upland and typical Dakota Country topography: a rather featureless plain developed on shales and smeared with recent sands and gravels. Dakota Country's western edge, however, is marked by a prominent, imposing, eastward-facing escarpment, or cuesta, capped with Greenhorn limestone.

The Blue Hills subdivision, stretching westward from this Greenhorn scarp, reaches to an even more prominent scarp—the Fort Hays chalk, which

WHENCE THE FENCE: GEOLOGY OF POST-ROCK COUNTRY

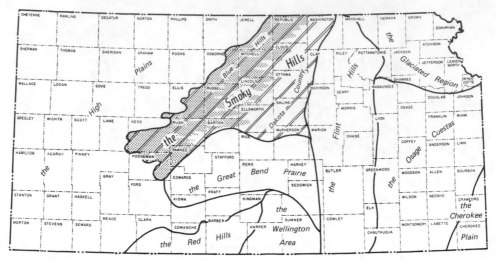

Physiographic divisions of Kansas. The shaded part of the Smoky Hills marks the Blue Hills subdivision. The eastern boundary, not well defined, generalized from Schoewe's 1949 map.

marks the eastern edge of the High Plains. Much more regular than the eastern Smoky Hills, the Blue Hills are flat-topped and tend to have prominent benches at various elevations on their slopes. The flat tops and the benches are found at similar elevations from one hill to the next, for they are held up by relatively resistant limestone beds that once were continuous over the entire area.

The landscape obviously is related to the rocks. So the rocks explain the difference in appearance between the eastern Smoky Hills and the Blue Hills.

First, the rocks underlying the eastern Smoky Hills are somewhat older than those of the Blue Hills. All except the youngest of central Kansas's sedimentary rocks dip gently (only a few feet to the mile) toward the west, whereas the land surface rises in that direction. So the farther west we go, the younger the rocks we see.

Second, much of the rock in the eastern Smoky Hills is quite different in origin from that in the Blue Hills. The variously colored clays and silt-stones and sandstones of the Dakota formation, well exposed in the eastern Smoky Hills, were deposited about a hundred million years ago on land, and they contain fossil remains of land plants, some of which are strikingly similar to modern land plants; magnolia, sassafras, fig, willow, and conifer, for example. The land was low-lying, flat, in places swampy; and streams meandered across it, leaving channel sands and floodplain silts and clays. As the deposits built up, the stream channels shifted; so relatively discontinuous sandstone beds now are found at various levels within the Dakota formation. Some of the vegetation in the swamp deposits became lignite or

Fort Hays escarpment, Osborne County (upper photograph). Perhaps the most conspicuous physiographic boundary in Kansas, the escarpment is highly visible as cliffs and ramparts in many places from eastern Jewell County southwest to northwestern Hodgeman County. Fencepost on the Blue Hills landscape, near Wilson Reservoir (middle photograph). Dakota Country's jumble of grass-covered hummocky hills and mounds, as seen near the east end of Wilson Reservoir in late afternoon (bottom photograph).

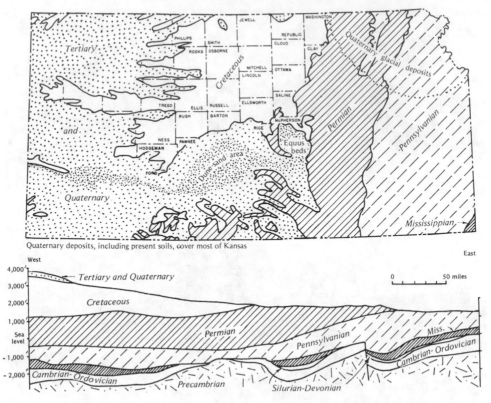

Generalized geologic map of Kansas and structural cross section.

brown coal, which much later was used by Kansas pioneers as fuel. Within the past million years or so the region has been dissected by modern erosion, and remnants of relatively resistant channel sandstone at various levels cap the hills. Some of the capping sandstone is hard and resistant because it is well cemented with dense, dark-brown iron oxide or limonite. Cross-beds, resulting from deposition from water currents, are common.

On the other hand, the rocks exposed in the Blue Hills are thin shales and chalks and limestones that were deposited in a broad, shallow sea that flooded over the Dakota deposits and at times extended across the middle of North America, all the way from Mexico to northern Alaska. It is sometimes called the Western Interior Sea. We can trace individual beds of limestone within the Greenhorn formation many miles across Kansas and into eastern Colorado, western Wyoming, and southeastern Montana, finding little variation in character or thickness. So when modern erosion dissected the Blue Hills, flat hilltops and benches developed on the most resistant of the limestones at specific but continuous horizons. One aspect of the Fencepost bed that made it so readily available is its presence at relatively shallow depths below the soil on extensive benches and flat hilltops.

THE ROCKS AND FOSSILS

The rocks of the stone-post country include, from east to west, the brightly colored clays, siltstones, and sandstones of the Dakota formation; a thin interval of dark-gray and olive-gray shale known as the Graneros shale; the Greenhorn limestone topped by the Fencepost bed; and at least the lower part of the Carlile shale.[1] These rocks were deposited during the Cretaceous Period—a time when dinosaurs reached their apex and started to decline, when plant life began to assume modern appearance, when flying reptiles lived on fish and are thought to have flown far out to sea. The shallow sea itself teemed with fish, clams, and myriads of microorganisms.

Perhaps one of the best places to see exposures of all four formations (Dakota, Graneros, Greenhorn, Carlile) is in road cuts around the Wilson Reservoir in Russell and Lincoln counties, in particular the north and south shore roads.

The Dakota rocks most readily observed in the eastern Smoky Hills are sandstones, which are light yellowish-brown to very dark brown, depending on the amount of iron oxide they contain, and which are commonly cross-bedded. These sandstones (mostly stream-channel deposits on the landward part of a gigantic delta) are particularly prominent where they cap hills. The claystones and clayey siltstones of the Dakota, much more abundant than the sandstones, are more likely to be grass-covered, but they are well exposed in many road cuts. These ancient floodplain materials vary from red to yellow to lavendar to gray to white; and color mottling, particularly red mottles in gray, silty claystone, is common. Fossil leaves or impressions of leaves have been found in many parts of the formation.

The upper fifty feet of the Dakota, however, is mostly dark-gray, carbonaceous shale and mudstone, with beds of lignite and a few beds of sandstone. The sediments were deposited near the shore of a sea that was gradually flooding northward and eastward over the region. The dark-gray (carbonaceous) shale and lignite are believed to have been the deposits in lagoons and marshes along the margin of the land. In a few places near the top of the formation, there are molds of marine shells, which signify the approach of the sea or perhaps infrequent storms that swept the shells landward.

The overlying Graneros shale (about thirty feet thick in the Wilson Reservoir area) is mostly dark-gray shale in its lower part; it contains a few thin beds of sandstone. The dark gray is caused in part by minute specks of carbonaceous plant material and in part by very fine-grained pyrite and marcasite, both of which are iron sulfide minerals. The fossils—particularly the brachiopods called *Lingula*—in the lower part represent the kinds of organisms that live in brackish water (less salty than normal seawater because it is mixed with fresh water from streams). In the upper part of the Graneros, however, are normal marine fossils, including ammonites, as well

Septarian concretion similar to those found in the Graneros shale. About one-half actual size. *Specimen provided by Myron Chapman, Beloit.*

as oyster shells and other fossil clams; and the shale is not quite so loaded with carbonaceous organic matter. Also in the upper Graneros are peculiar structures known as septarian concretions, which consist largely of "septa" of yellow-to-brown calcite enclosing some of the surrounding sediment.

Some of the Graneros sandstone has a yellow stain produced by the mineral jarosite, a hydrous potassium iron sulfate. Gypsum crystals also are present in some of the sandstone and silty shale.

The Graneros shale contains several layers of bentonitic clay. Most are less than four inches thick, though one bed near the top of the formation is unusually thick, averaging about a foot. Bentonite, a soft, unctuous clay when wet, brittle and much fractured when dry, is altered volcanic ash. This particular one-foot bed has been traced in the subsurface into western Colorado and also has been recognized in Wyoming and Montana. It thickens toward the west, suggesting that the volcanic source was somewhere west of the western edge of the Western Interior Sea.

The Greenhorn is a series of thinly laminated beds of shaly chalk, chalk, chalky limestone, and bentonite and is approximately ninety-five to one hundred feet thick in post-rock country. The Greenhorn has been subdivided into four members, from bottom to top: Lincoln limestone (about twenty to twenty-five feet thick), Hartland shale (thirty to thirty-five feet), Jetmore chalk (twenty to twenty-two feet), and Pfeifer shale (twenty to twenty-two feet, including the eight- to twelve-inch Fencepost bed at the top).

On casual observation, all Greenhorn members seem to be alike. Most exposures show thin, rather grayish-orange to yellowish-gray, thinly laminated beds of alternating soft shaly chalk and slightly more resistant chalk, along with chalky limestone. Clams abound throughout, along with a few

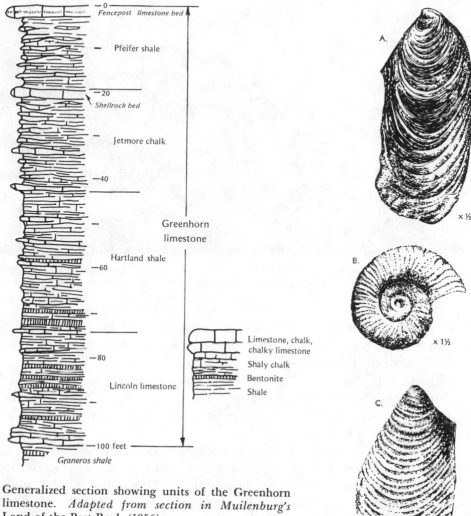

Generalized section showing units of the Greenhorn limestone. *Adapted from section in Muilenburg's* Land of the Post Rock *(1956).*

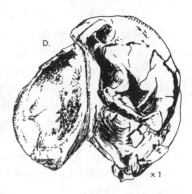

A few inhabitants of the Greenhorn sea. A) *Inoceramus labiatus,* abundant in Jetmore and Pfeifer rocks and also found in the lowermost part of the Fairport chalk member of the Carlile shale; B) *Watinoceras reesidei,* an ammonite from the Jetmore chalk member; C) *Inoceramus pictus,* a pelecypod found in the Hartland and Lincoln members; D) *Ostrea beloiti,* in the upper part of the Graneros shale as well as in the lower part of the Lincoln member of the Greenhorn.

Upper part of Greenhorn formation, about three miles north of Bunker Hill. This fine roadcut is capped by the Fencepost bed, which in turn is crowned by the Fencepost limestone fence posts.

worm burrows, ammonites, fish remains (especially sharks' teeth), and, in the lower part, oysters. By far the most common clam is *Inoceramus*. "Worm-"bored driftwood is also found.

When freshly exposed, Greenhorn sediments are relatively dark and have an olive-gray cast, which may be attributed to abundant organic matter and to finely disseminated pyrite or marcasite; but after the rock is exposed to the air the color lightens.

The Lincoln limestone member has a cross-bedded, dense limestone at its base, overlain by thin beds of shaly chalk and chalky limestone and numerous thin seams of bentonite. Oyster shells *(Ostrea beloiti)* and shark teeth are common at the base, along with the pelecypod *Inoceramus pictus*. Thin beds of hard, chalky limestone near the top of the Lincoln member probably are the source of much of the flagstone that has been used throughout post-rock country.

The Hartland shale member is not strictly shale, but rather predominantly shaly chalk with a few scattered beds of chalk and chalky limestone and a few bentonite seams.

The Jetmore chalk member, a cliff former, contains several relatively thick beds of chalky limestone—particularly the Shellrock bed at the top—and thinner shaly chalk beds.

The Pfeifer shale member consists of shaly chalk and chalky limestone, with a few bentonites; it should not be called shale. It differs from the Jetmore in that the Pfeifer has more nodules and beds of hard chalky limestone. Many *Inoceramus labiatus* fossils can be found in the shaly chalk of the lower two-thirds.

Inoceramus shells (chiefly I. labiatus) on upper surface of Shellrock bed in a Cloud County quarry.

"Sugar sand" bentonite six feet below the Fencepost limestone. Generally, it must be dug out to be seen.

Doubtless of greatest interest to post-rock buffs is the upper part of the Greenhorn with its crowning glory, the Fencepost bed. So let us take a closer look at the Pfeifer member.

The Pfeifer overlies a relatively thick (twelve inches or more), prominent, bench-forming limestone that is so loaded with *Inoceramus labiatus* and a few other fossils that it is known as the "shell bed," "*Inoceramus* bed," or "Shellrock." Because it is relatively resistant and a bench former it has been used to a limited extent for fence posts and building material.

The basal fourteen feet of the Pfeifer consists of thin beds of shaly chalk and chalky limestone with numerous fossils (chiefly *Inoceramus labiatus*). One of the thin, chalky limestone beds, about nine feet below the Fencepost limestone, is a relatively resistant iron-stained limestone somewhat similar to the Fencepost but only five inches thick.

Another well-known bed, long described as "sugar sand," is a four-inch bentonite that includes a layer of tiny, white calcite or gypsum crystals. The bentonitic clay is dark yellowish-orange, occurs about six feet below the Fencepost limestone, and is similar in character to the other bentonites in the Greenhorn and lower Fairport. Rubey and Bass in 1925 pointed out that the bentonites "are easily recognizable by a peculiar jointing that has been appropriately likened to the texture of fish meat" (*The Geology of Russell County, Kansas*).

Nodules or rounded concretions of dense, hard limestone are particularly abundant in layers within a few feet of the Fencepost bed—both above and below. Similar nodules also occur at various other positions within the Greenhorn. Some have *Inoceramus* shells at the top or inside. The nodules generally are twelve to eighteen inches in diameter and four to ten inches thick. Most of them show thin lamination. A photomicrograph would show layers consisting of tiny, prismatic fragments of *Inoceramus* shell, sandwiched between layers made up of shells of one-celled organisms called *Globigerina* (a common fossil in chalk). This separation seems to be the result of sorting by gentle wave action on the sea floor, the finer-grained calcareous ooze having been winnowed out. Many of the hard limestones of the upper Greenhorn consist predominantly of *Inoceramus* shell prisms. Reeside suggests that the shells were chewed up by shell-crushing sharks (*Paleoecology of the Cretaceous Seas of the Western Interior of the United States*).

The hard nodular limestone and perhaps other dense, hard Greenhorn limestones have been used, particularly in Lincoln County, for decorative purposes. The limestone can be sawed and polished and is known in Lincoln County as "Lincoln marble." Perhaps a better term would be "crystalline limestone."

The Fencepost bed itself is a remarkably persistent, uniform bed, eight to twelve inches thick. Generally its thickness is nine inches or less. Its brown streak (limonite stain) is especially well developed in the northeastern part of the area. Fossils are few compared with the Shellrock bed, but include *Inoceramus labiatus, Collignoceras woolgari* (an ammonite), fish scales, shark teeth, and driftwood. The limestone is hard compared with the shaly chalk beds, and Hattin suggests that this and other such resistant beds may have originated during wave or current stirring of carbonate mud on the sea floor (*Upper Cretaceous Stratigraphy, Paleontology, and Paleoecology of Western Kansas*). Such an origin may possibly account for the uneven or irregular upper and lower surfaces of the Fencepost limestone; both surfaces suggest slight wave erosion.

Some of the other hard limestones, according to Hattin, resulted from recrystallization of fine-grained carbonate ooze.

The Fencepost limestone is unusually free from joints, or linear breaks. Most limestones and other solid rock masses are crisscrossed by sets of joints, often two sets almost at right angles to each other. In fact, joints are said to be the most characteristic feature of rocks exposed at the earth's surface. A reasonable explanation of joints is simply that they are produced by release of pressure. Once buried and subjected to pressure from overlying rocks, but now at the surface, the rocks tend to break apart, especially as a result of slight movement of the earth's crust. We found no evidence of joint sets in

Hard limestone nodules in chalky shale below Fencepost limestone in roadcut two miles north of Bunker Hill (left-hand photograph) and limestone nodules exposed in road ditch near entrance to Minooka Park. Rainwater has worn away the surrounding soft, shaly chalk.

the Fencepost bed, and such directions of breakage that do exist seem more related to the slope of the local landscape than to any specific directions of regional jointing. We have no facile explanation for the joint-free character of the post rock, but chalky limestones the world over generally contain few joints.

The Fencepost bed marks the top of the Greenhorn formation, but the beds in the lower twenty-five feet of the overlying Fairport member of the Carlile shale are so similar to the Pfeifer rocks that perhaps they should have been a part of the Greenhorn. The Fencepost bed seems to have been chosen as the top of the Greenhorn for two reasons: (1) It is easily traced and mapped, and (2) early workers believed that *Inoceramus labiatus* was restricted to the Greenhorn and could not be found above the Fencepost bed (Rubey and Bass, *Geology of Russell County*). More recently, however, geologists have found that *I. labiatus* is relatively abundant in the lower part of the Fairport.

Along with thin-bedded, shaly chalk in the five-foot interval of Fairport chalk directly above the Fencepost bed we find numerous limestone nodules, two thin seams of bentonite (one near the base and the other about five feet up), and a fairly prominent resistant bed of yellowish-gray, chalky limestone reminiscent of Fencepost limestone. This limestone may be the source of much of the stone brick.

THE BROWN STREAK (S)

Brown streaks are rare in limestones. Perhaps that is one reason the brown streak "near the center of the bed" has been noted by everyone who has recorded descriptions of the Fencepost limestone bed. It is known to be a form of "iron stain"—limonite, if you will, or, more accurately, hydrogen iron oxide or goethite ($HFeO_2$) similar to ordinary brown iron rust. A few geologists and others have also noted that several other limestone beds and layers of nodules in the Greenhorn limestone contain more-or-less central brown streaks.

The iron probably is derived from alteration of finely disseminated pyrite and then redistributed during a process of recrystallizing and cementing of the limestone, followed by oxidation when the limestone is exposed to the air. But why is the brown streak so rare in limestones?

Perhaps certain features that the beds containing brown streaks have in common may provide clues to their origin.

First, what about the Fencepost bed itself? Despite tradition, the brown streak is not really "central." Where best developed it lies somewhat below the center of the bed. But Hattin, no doubt facetiously, bowed to tradition when he referred to it as in the "*center* of [the] lower 0.7 foot" of the bed (emphasis added). Other astute observers (not primarily geologists, incidentally) have noted that the prominent brown streak near the center of the bed is found only in the northeastern part of post-rock country. As we trace the Fencepost bed to the southwest, we can see that the brown streak becomes paler and less "central." Farther west and southwest it is replaced by two or even three (rarely as many as six) much paler and thinner brown streaks, none of which is central. In the southernmost part of the land of the post rock the streaks are very poorly developed, and the rest of the bed is much paler than the yellowish Fencepost limestone to the northeast (see color plates). How can we account for this variation?

Occurrences of a brown streak in chalky limestone beds of the Greenhorn formation may provide a reasonable explanation. The streaks are common in such beds in the lower few feet of the Pfeifer. Two chalky limestones in the Jetmore member contain brown streaks; and there is one near the base of the Lincoln member. These beds, not quite so spectacular as the Fencepost bed, are relatively harder than most Greenhorn ledges, and, like the Fencepost, they contain fossil shells. Another brown streak may be seen in a nodular zone above the Fencepost bed, traversing the centers of the nodules, which also bear fossil shells.

Unfortunately none of the hard statistical facts are available, but our tentative impression is that, within the Pfeifer, more specimens of land-derived fossil wood have been found at the brown-streaked horizons in the northern part of the post-rock area than in the southern part.

Resistant beds, driftwood, numerous fossil shells, regionally variable brown streaks—what is the significance of their association? The softer chalk beds are shaly and made up of impure, very fine-grained lime mud and microfossils. But the harder limestones contain larger units of calcite, some of which were formed as the finer-grained calcite recrystallized. Donald Hattin's classic study of the Carlile shale, published in 1962, suggests that the hard limestones and nodules of the basal Carlile (Fairport member) were formed by deep-reaching waves that reworked the finer sediment and produced layers susceptible to recrystallization. Wave action would also tend to concentrate the larger shell fragments and winnow out the fine lime mud. Hattin's explanation would apply equally well to at least some of the hard, shell-bearing limestones of the Greenhorn.

In 1967 Sanders and Friedman ("Origin and Occurrence of Limestones"), describing the chalk of northwestern Europe, noted that in the London–Paris Basin hard chalks are most abundant around the basin margins (which of course would have been areas of shallower water and consequently greater wave action).

We need more information to explain the regional variations within the Fencepost bed and the association of iron stain (brown streak) with hard chalk. Geochemists have noted that the iron content of sea water is extremely low—in the range of 0.002 to 0.020 part per million, whereas the iron in

Sawed section of driftwood found just above the Fencepost bed. The log is encased in a hard calcareous shell, forming an elongate concretion. The wood itself has not been replaced; it could be thought of as a mini-lignite deposit. The rounded areas are calcite, which fills holes made by organisms sometimes referred to as marine "worms" (*Teredo,* or sometimes Parapholas). The animals were molluscs similar to the modern "shipworm." Shown at about three-fifths actual size. *Specimen provided by Harvey Roush.* (Some of the fossil wood from the Greenhorn is replaced by quartz and contains small, sparkling crystals. A sample of this silicified wood was on display at the Post Rock Museum in late 1973.)

river water averages about 0.5 part per million. Lagoon, bog, and swamp waters seem to be particularly rich in iron. It seems reasonable to expect that storm waves roiling up the bottom of the Cretaceous sea might be associated with torrential rains bringing floods of iron-rich fresh water from the low, swampy land, along with tree trunks and other debris.

The precise distance of the eastern shoreline from what is now post-rock country can never be known because erosion has removed the Fencepost bed to the east. However, a map compiled by J. B. Reeside, Jr., one of the foremost experts on Cretaceous rocks and history, depicts his concepts of the distribution of land and sea during late Greenhorn time, based in part on

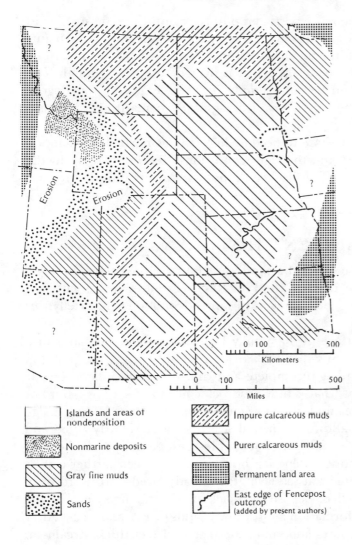

Paleogeographic map showing sediments and inferred land area during late Greenhorn time. *Adapted from "Palcocology of the Cretaccous Seas of the Western Interior of the United States," by J. B. Reeside, Jr., 1957, and printed by permission of the Geological Society of America.*

Islands and areas of nondeposition

Nonmarine deposits

Gray fine muds

Sands

Impure calcareous muds

Purer calcareous muds

Permanent land area

East edge of Fencepost outcrop (added by present authors)

the distribution of the various types of sediments that were being deposited. In his map, the inferred shore is not far east of post-rock country. It is reasonable to suggest that the northeastern part of post-rock country was closer to land than the southwestern part was, thus providing an explanation for the richer iron streak to the northeast. The land itself, as it had been in Dakota time, must have been low-lying and swampy, for it contributed very little sand or silt to the limestone. One can imagine a major river system, similar to the modern Mississippi but carrying much less silt, emptying into the Cretaceous sea from the northeast and transporting significant quantities of iron compounds (and some driftwood) far offshore during and after a torrential storm; and farther south, perhaps, smaller streams having a lesser effect. Such hypothesizing is enjoyable, but that is all it is, because other evidence is scanty or absent.

The red streaks of the fence posts and post-rock buildings are such vivid reminders of prairie fires and urban disasters that they deserve an explanation. Limonite or goethite, the brown-streak mineral, can be described roughly as a hydrous iron oxide. When it is heated to a high enough temperature, it loses water and becomes simply iron oxide (Fe_2O_3), analogous to hematite. The characteristic color of fine-grained hematite is brownish-red.

Goethite also can dehydrate to become hematite at normal low temperatures, given enough time and the right conditions. The red streak in the Mehl quarry is not a product of fire; we cannot explain its unique presence in the limited area.

ANCIENT HISTORY

Putting it all together, we can add a significant fourth dimension—geologic time—to our story of the Land of the Post Rock. From evidence garnered by several generations of geologists working in the Smoky Hills, briefly this is the way it looks.

The rocks at the surface in the Smoky Hills are meager remnants of a long series of geologic events that took place roughly within the past 130 million years. The record is incomplete and difficult to decipher. As our history begins, we see that what is now post-rock country was a low—but not particularly flat—land area underlain by gently dipping rocks of Permian age deposited in and near shallow continental seas and later slightly tilted and eroded. Permian rocks generally are not visible in the Smoky Hills because they are buried under younger deposits; but they can be seen farther east.

On this low land, almost at sea level, the sea advanced northward from what is now central Mexico, depositing black muds and fine sand of the Early Cretaceous Kiowa formation amidst low islands of Permian rock. After the sea retreated, leaving mud and sand several tens of feet thick, streams in Dakota time flowing from the northeast built up about four hundred feet of deltaic deposits: sand, silt, and clay, with swamp deposits rich in vegetation (now lignite).

Then again the sea crept into Kansas from the south, depositing dark-gray mud and some sand—the Graneros formation. The thick bentonite at the top records an important volcanic eruption somewhere to the west of Kansas. There seems to have been a brief interval of erosion before basal Greenhorn limestones were deposited.

In early Greenhorn time (Lincoln and Hartland members), the sea deepened and spread from the west, apparently all the way to northern Alaska and the Arctic regions, for northern varieties of fossils are found in the Kansas Greenhorn rocks. The country west of the western shoreline was the site of mountains and volcanoes. The relative purity of the chalky sedi-

ERAS	PERIODS	ESTIMATED LENGTH IN YEARS*	TYPE OF ROCK IN KANSAS	PRINCIPAL MINERAL RESOURCES
CENOZOIC	QUATERNARY (PLEISTOCENE)	1,000,000	Glacial drift; river silt, sand, and gravel; dune sand; wind-blown silt (loess); volcanic ash.	Water, agricultural soils, sand and gravel, volcanic ash.
CENOZOIC	TERTIARY	59,000,000	River silt, sand, and gravel; fresh-water limestone; volcanic ash; bentonite; diatomaceous marl; opaline sandstone.	Water, sand and gravel, volcanic ash, diatomaceous marl.
MESOZOIC	CRETACEOUS	70,000,000	Chalk, chalky shale, dark shale, vari-colored clay, sandstone, conglomerate Outcropping igneous rock.	Ceramic materials; building stone, concrete aggregate, and other construction rock; water.
MESOZOIC	JURASSIC	25,000,000	Sandstones and shales, chiefly subsurface.	
MESOZOIC	TRIASSIC	30,000,000	Sandstones and shales, chiefly subsurface.	
PALEOZOIC	PERMIAN	25,000,000	Limestone; shale; evaporites (salt, gypsum, anhydrite); red sandstone and siltstone; chert; some dolomite.	Natural gas; salt; gypsum; building stone, concrete aggregate, and other construction materials; water.
PALEOZOIC	PENNSYLVANIAN	25,000,000	Alternating marine and non-marine shale, limestone, and sandstone; coal; chert.	Oil, coal, limestone and shale for cement manufacture, ceramic materials, construction rock, agricultural lime, gas, water.
PALEOZOIC	MISSISSIPPIAN	30,000,000	Mostly limestone, predominantly cherty.	Oil, zinc, lead, gas, chat and other construction materials.
PALEOZOIC	DEVONIAN	55,000,000	Subsurface only. Limestone, black shale.	Oil
PALEOZOIC	SILURIAN	40,000,000	Subsurface only. Limestone.	Oil
PALEOZOIC	ORDOVICIAN	80,000,000	Subsurface only. Limestone, dolomite, sandstone, shale.	Oil, gas, water.
PALEOZOIC	CAMBRIAN	80,000,000	Subsurface only. Dolomite, sandstone.	Oil
PRE-CAMBRIAN	(Including PROTEROZOIC and ARCHEOZOIC ERAS)	1,600,000,000 +	Subsurface only. Granite, other igneous rocks, and metamorphic rocks.	Oil and gas.

(Not scaled for geologic time or thickness of deposits)

*Committee on Measurement of Geologic Time, National Research Council

State Geological Survey of Kansas

Geologic timetable and rock chart. *Reproduced from* Kansas Rocks and Minerals *by Tolsted and Swineford, 1957.* (Since the timetable was first published, slight modifications have been made in estimated durations of the various periods.)

ments in central Kansas records very little input of mud from the nearby eastern shore; thus, country to the east must have been extremely low and nearly flat. Western volcanoes contributed volcanic ash, which altered to bentonitic clay in the Greenhorn sea. Toward the end of Greenhorn deposition, however, the sea may have shrunk somewhat, the eastern shoreline (or at least the area of calcareous muds) creeping slightly basinward. The hard chalky limestones (such as the Fencepost bed) of the Greenhorn, described by Hattin as resulting from periods of wave action and suggesting relatively shallow water, became more abundant in late Greenhorn and early Fairport time.

The sea continued to shift or shrink during and after Fairport time. There may have been a period of erosion when central Kansas was not under water. The deposition of Fort Hays chalk marked the beginning of the greatest expansion of the sea in Cretaceous time; sea water apparently covered almost one-third of North America.

In post-rock country there is no record of events immediately after deposition of the Fort Hays chalk, because younger Cretaceous rocks that undoubtedly were once there have been eroded away. But at some time near the close of the Cretaceous Period, about seventy million years ago, the seas left Kansas, and the Rocky Mountains were born. By Pliocene time (about ten million years ago), sand and gravel carried by streams from the Rockies built up the High Plains and reached into the Smoky Hills area; scattered remnants are still to be found at a few places in the Smoky Hills.

During the past one or two million years (the Pleistocene or Ice Age) up to the present, the main event has been erosion, principally by streams. There are, however, some deposits of Pleistocene age. Climatic changes and the presence of glaciers elsewhere resulted in deposition of silt and sand in broad floodplains of major streams and outwash channels from the melting glacial ice. Associated with these deposits are areas of dune sand and scattered lenticular beds of volcanic ash and a mantle of wind-blown silt (loess) that was picked up by the wind from the floodplains of streams that carried glacial meltwater.

It is erosion, however, that is responsible for sculpturing the Smoky Hills, for developing the escarpments of Greenhorn limestone and the broad benches underlain by the Fencepost bed. A million years of attack by rainfall, gravity, and meandering streams has produced the landscapes of today.

Scene in southern Mitchell County.

The frontiers are not east or west, north or south, but wherever a man fronts a fact.
—Thoreau

Cook house, northwestern Lincoln County.

Little I ask; my wants are few,
I only wish a hut of stone,
(A very plain brown stone will do,)
That I may call my own.

—Oliver Wendell Holmes

Russell County
Ness County

Lincoln County
Rush County

Mitchell County
Barton County

Stone posts, showing variation in brown streaks and coloring,
in several post-rock counties.

Remains of ranch house, western Lincoln
County, built before the railroad reached
Lincoln.

*The walls show . . . a strength that is al-
most defiant.*

Storm in late afternoon near Paradise,
Russell County.

*Stone posts . . . shining, like elongated
lights on a landscape . . . soon to be covered
by the night.*

6

ON A JOURNEY
THROUGH THE LAND OF
THE POST ROCK

Linked as it is with human occupancy, the Land of the Post Rock records regional history and economic development. Where cattle graze behind well-kept fences supported by stone posts, there is continuity in the use of post rock to maintain control over almost treeless farmland. Where stone posts set years ago are still standing but with their wires in disarray and obviously serving no purpose, there has been some break in the use of the resource or of the land. Likewise, along many roadways stone houses, some lived in and some not, reveal by their construction or architecture or state of repair something about the homesteaders or their descendants and their success or failure in living off their quarter sections of land. Post rock in some form, as posts or as building stone or as ledges along the escarpments, is obviously enough to command the attention of and inform anyone who journeys along or across the divides of this Kansas plains borderland, drained by the Republican, the Solomon, the Saline, the Smoky Hill, and a few tributaries of the Arkansas.

UP THE REPUBLICAN, ABOVE THE SOLOMON

For geographic and chronological continuity, it is convenient to begin a journey in post-rock country by entering it from the northeast, moving gradually southwestward, guided by a good road map of Kansas. The first natural subdivision to explore, then, would lie north of the Solomon River

Along Kansas 14, Lincoln County.

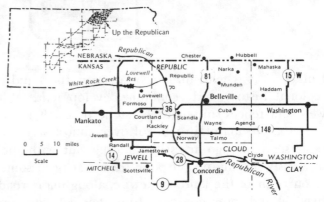

Land of the Post Rock (shaded area on map) can be divided into three major divisions according to drainage: area drained by 1) the Republican River; 2) the Smoky Hill River and its major tributaries, the Solomon and the Saline; and 3) the Arkansas.

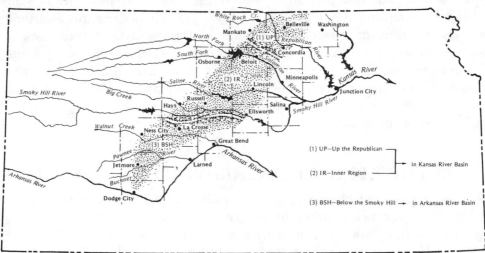

(1) UP—Up the Republican ⎤
⎟ in Kansas River Basin
(2) IR—Inner Region ⎦

(3) BSH—Below the Smoky Hill → in Arkansas River Basin

between Mankato and Washington and would include the highlands along the Republican, where Scandinavians, Englishmen and Scotsmen, and Bohemians began quarrying native limestone to establish permanent settlements even before 1870.

Venturing up the Republican, as some of the Swedes and Britons did soon after they came to Republic County, one comes to the site of White Rock, once a flourishing trade center where White Rock Creek leaves Republic County to enter Jewell. Only one residence, a house partly of stone on the Jewell County side, remains to mark this town the railroad bypassed. The stone part was built in 1869 and was the town's first permanent dwelling, according to Mrs. Winifred A. Smies of Courtland, who once lived on this townsite farm, which she now owns and which is lived on and operated by her son William and his family. Mrs. Smies added that Mr. and Mrs. Allen D. Woodruff, who homesteaded the place, hauled rock for their house on a sled pulled by an ox team from a quarry (on a farm now owned by Mrs. Jennie Blanding) about five miles southwest of the home site. Though the rock in the house is Fort Hays chalk, softer and more massive than Fencepost limestone, the house can be considered a boundary landmark. Post rock has been used extensively to the south and east.[1]

First house in and only house remaining at site of White Rock. The stone (middle) part was the original house.

Stone skills of the Bohemians are evident particularly in eastern Republic County, from Cuba north to Munden and Narka and south to Agenda, and in adjacent Washington County, at least as far east as Haddam. The evidences, however, do not include many stone posts, few of which were used in those localities, where rainfall generally assured the presence of enough timber for posts. But much rock was quarried for building stone, as we are reminded by the old quarries along U.S. 36 between Haddam and Belleville. (Except for western and northwestern Washington County, all of post-rock country lies west of the sixth principal meridian, which divides east and west ranges and coincidentally lies along the Washington-Republic county line.)

In Narka we observed a few footings and a few small culverts constructed of Fencepost limestone, but the northernmost building of Fencepost limestone we saw was on the J. B. Rickard homestead (now owned by Ben Blecha of Munden), about three and a half miles north of Munden, near the turn in the paved road. The walls of the two-story house are of rough-hewn, rectangular blocks of various lengths but of relatively uniform height. The wide brown stripes through the centers of the deep-tan blocks, which have a smooth finish, give the now abandoned but structurally intact house a distinctive appearance. Architecturally simple, it suits the regional environment. When constructed in 1885 by Rickard, it must have been the envy of area residents. We consider it a post-rock landmark of the northeastern part of the region.

About four miles north of Cuba, on property now owned by Mr. and Mrs. Eddie Popelka and occupied by the Ralph Wests, we examined a two-story, twelve-room house of native limestone built over a cellar and two stone-arch caves. The light-buff stone, with pale-brown streaks, probably was quarried from one of the limestone beds associated with the Fencepost. The late Edward W. Shulda of Belleville, who once lived on the farm, said that Joe Baxa, a stonemason, and a carpenter named Hulka built the house for the J. J. Shimek family in 1903. Bessie (Mrs. George) Lang, nonagenarian who has lived most of her life in the neighborhood, said that both Baxa and Hulka were educated and learned their trades in Czechoslovakia.

Among stone buildings in present Cuba (the town was relocated along the railroad line in 1884), the stone blacksmith shop built about 1890 by Mitchell Davidson is now somewhat a curiosity. The original forge used by Davidson and later by Frank Sterba remains in place, and an anvil and old tools may be seen. True to the sign outside, Joe Sterba, a nephew of Frank Sterba, continues to do blacksmithing there, using electricity to heat the forge.

The early residents of Belleville, about thirty miles west of Cuba, were mostly English-speaking people born in the United States. They may have been less inclined to build with stone than were the newer European ar-

Rickard house, north of
Munden.

Shimek House, north of
Cuba.

Cuba's blacksmith shop for more than eighty years.

House built in 1877 by a Scottish stonemason; located near US 81 about four miles south of Belleville, Republic County.

rivals. Yet an article in the June 10, 1890, issue of the *Topeka Daily Capital* emphasized Belleville's attention to erecting permanent structures near the beginning of the building-stone age: "The business blocks are substantially built of brick and native stone, the frame buildings of the pioneer days having disappeared."

One of Belleville's business buildings, erected in the mid 1880s as the First National Bank, is a blend of several building stones, including some Fencepost limestone and some sandstone (probably Dakota), but predominantly limestone of an older geologic age (Permian). The two-story, somewhat ornate building is still standing and in use, apparently still structurally sound. A Scottish stonemason, James Doctor (brother of Robert, Tom, and Peter mentioned in the next paragraph), worked on that bank, according to a nephew, Ben Doctor of Lawrence. Mrs. George Doctor of Belleville pointed out that James also did the stonework on a bank in nearby Scandia and on that bank carved the Scottish emblem: a thistle.

A few stone buildings of 1870–1880 vintage keep history alive on the Scotch Plains southwest of Belleville and link community development to the coming of Scottish immigrants as members of one division of the Ex-

Section of wall of barn on the Peter Doctor homestead on the Scotch Plains, Republic County, showing how that early stonemason "finished" the limestone blocks and laid them to reveal the wide, brown streak.

Unusual hitching ring for horses; in wall of barn built by Peter Doctor.

celsior Colony (the second division in Republic County, but the third if we count the one that attempted to settle along White Rock Creek). Notable are the stone structures built by members of the Alexander Doctor family; four of five brothers (all sons of Alexander) were stonemasons. Stonework of Robert, Tom, and Peter may be seen on the three adjoining quarter sections they homesteaded, and a stone house along U.S. 81 about four miles south of Belleville still stands on property once owned by Alexander Doctor (now owned by George Sis of Belleville). A stone house on the Robert Doctor farm, now owned by Lawrence Dooley, is lived in by the Melvern Cornetts. According to a son, the late George Doctor of Belleville, the stone was quarried on the place. The house (eighteen by thirty feet) and a stone barn (twenty-two by fifty feet) are mentioned in Andreas's *History of the State of Kansas*, which also records that Robert Doctor "has stone enough on his place to last him a lifetime, of good quality and of uniform thickness." Loren Howe of Belleville, who owns the Tom Doctor homestead north across the road, uses the stone house on that place as his headquarters on work visits to the farm. A barn on the Peter Doctor homestead, east across West Creek, is still in use; the property is owned by Frances Walter Elyea and son Larry Walter.

Remains of barn wall and house on the A. S. Johnson place south of Courtland, Republic County. Property now owned by Mrs. J. Rudolph Johnson, a granddaughter-in-law of A. S. Johnson.

Some members of the Excelsior Colony, including Alexander Doctor's son-in-law Alexander Henderson, a stonemason, settled farther south, along the Republic-Cloud county line in the vicinity of Minersville, which for a time became a flourishing coal-mining community.[2] The Scottish craftsmen obtained their building material from both Fencepost and Shellrock beds of the Greenhorn limestone in nearby bluffs, but those who settled in Cloud County used chiefly Shellrock for their houses and for fence posts as well.

Handiwork of the Scandinavian stonemasons is especially evident in the vicinity of Scandia and in the Kackley and Norway communities of southwestern Republic County. In Scandia we visited with Dr. C. V. Haggman, an alert ninety-six in 1972. He told us that many of the area's early stonemasons learned their trade under the apprentice system in the old countries and had become master workmen before arriving in this country. He mentioned as one example the Andrew S. Johnson farm about three and a half miles south of Courtland. That farm, with its stone buildings, once was considered a showplace. The house and one wall of the barn, built of Fencepost limestone in 1872, are still standing. The rectangular blocks were quarried and shaped with hand tools and fitted in a rough-hewn fashion into walls two feet thick. The smooth, curved stone trim over the windows and

"Dream mansion" of post rock south of Scandia. Built in the early 1890s by Thure Wohlfort, one of the original colonists of New Scandinavia (Scandia), it replaced a smaller stone house, which in turn replaced a dugout. *From an old print possessed by Herbert Sandell.*

doors, shaped from limestone slabs so that a lighter face would show, contrasts with the tan, brown-streaked limestone in the barn's walls.

Some of the Scandinavian settlers first built simple stone houses and then as soon as they began to prosper built more commodious stone homes. An outstanding example of a mansion is the one built by Thure Wohlfort in 1893 on his homestead about a mile and a half south of Scandia. Wohlfort built a huge barn on his land in 1885. The mansion, the barn, and other stone buildings, including the original stone house built in 1870, are being kept in good repair by Wohlfort's grandson and current owner, Herbert Sandell, Manhattan lawyer, and by ranch operator Edward L. (Eddie) Kunze.

The best preserved and most prominent stone structure in southwestern Republic County is the Ada Lutheran Church, with its unusually tall spire. (Dr. Haggman said he often has heard the church's "extremely goodtoned bell" in Scandia, more than three miles distant.) The main part of the church was completed in 1884 under the supervision of Andrew Hedstrom, who received $200 for his services. Other church members donated labor and teams for hauling rock from the quarry, three miles west, near the

Republic-Jewell county line. Feathers and wedges were used to quarry the rock, Fencepost limestone. According to Andrew Hedstrom's son, Ephriam Hedstrom, who lives near the church and has served as sexton for more than sixty-five years, "men of the congregation hand sawed the blocks so they would be smooth, then my father made one face rough with his stone hammer. . . . He laid the outside walls and another man fitted rocks from the inside to make the two-foot thick walls. . . . Each rock was laid with unslaked lime for mortar."

To lay the rocks in the upper portion of the walls, scaffolds were used, stone being passed by hand from one level to the next. In recent years, the church has been remodeled and a wing added, but it is the simple artistry of the early stonework that gives it strength to match its years.

Ada Lutheran Church, located near Kackley in Republic County.

About one and a half miles east and five and a half miles south of the Ada Lutheran Church, in the Jamestown community in Cloud County, we can observe the stone church built by Danish Baptists who settled there. The inscription over the arched door reads: Church of Saron, Skandinavisk Baptist Kirke 1877. The architecture of the rectangular church (twenty-four by thirty-three feet) of rough-hewn Shellrock is simple. According to Burnice Pearson (Mrs. George) Jensen, Peter Chrestenson was hired to do the masonry work. Much labor, however, was donated by church members, some of whom quarried and hauled rock from about one mile northeast of the church. Total cost was in the neighborhood of $1,000. The little church and adjoining cemetery have been well cared for. Mrs. Jensen said that the church was redecorated for its hundredth anniversary (July 25, 1971).

At Concordia, near where the waters of the Republican leave post-rock country, one of the most impressive edifices of native limestone is the Catholic church, "Our Lady of Perpetual Help," built in 1902 under the direction of French Canadians who settled in that vicinity in the 1870s. Although most of the rectangular blocks, of both Fencepost and Shellrock, are laid horizontally in even-height courses in the church's walls, some appear in vertical position (as "shiners") for decoration or contrast. Except for the cut stone trim, including corner quoins and headers for windows and doors, the exposed faces of the blocks are rough and have hammer marks showing. Depending on the face exposed, the limestone exhibits different weathering patterns. The shiners, particularly the Shellrock shiners, have weathered the most.

Shellrock has been used more extensively than Fencepost in business buildings in Concordia. The public library, erected in 1908, is of Shellrock; except for the stone trim on its front, so is the stone part of the Brown Grand, opened as an opera house in 1907 and now a theater. In the country-side north and west of town, fence posts, as well as buildings, of Shellrock can be recognized. The posts have odd shapes and numerous fossils. But westward and southward toward the Solomon Valley, post rock, which (being stratigraphically higher) crops out west of the Shellrock, resumes its dominance as a material for buildings and posts.

THE INNER REGION: FROM THE SOLOMON TO THE SALINE TO THE SMOKY HILL

Stone buildings and rows of stone fence posts appear with such regularity in post-rock country's heartland that they can scarcely escape the notice of even the traveler who keeps his eyes on the road. This inner region is outlined on the north by the escarpments on either side of the Solomon from Delphos in northwestern Ottawa County to Downs in Osborne County. It extends southward across Mitchell and Lincoln counties into Ellsworth County as

ON A JOURNEY THROUGH THE LAND OF THE POST ROCK

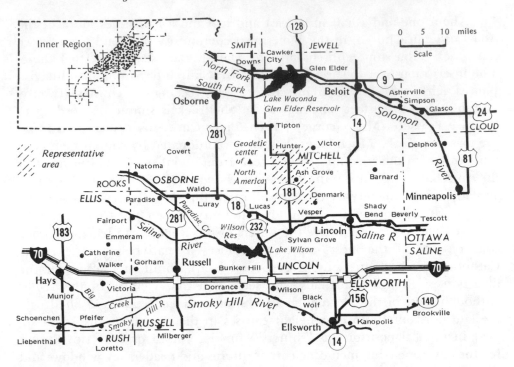

far as Ellsworth. It extends westward along the Saline and Smoky Hill valleys across Russell, southern Osborne, and northern Barton counties into Ellis County almost to Hays, and it includes a portion of northern Rush County as far west as Liebenthal.

The post rock of this broad area is a somewhat creamier buff than that of the northeastern extension of the region. Though generally distinct, brown streaks in the post-rock landmarks here appear narrower and somewhat paler than those in the landmarks in the northeastern part of the region. Also, the streak is not everywhere centrally located and there may be more than one. The paler color and multiple streaks become progressively more noticeable toward the southwest. (Contrasts are obvious in the photographs of post-rock posts, from various parts of the region, in the color plates.)

Brown streaks and color aside, here in the inner reaches is where the post-rock motif closely links the limestone resource and the history of regional development. Here, the escarpments along the Solomon and the Saline and the Smoky Hill commonly show off their pale-buff limestone ledges within sight of, occasionally immediately below, rows of stone posts. The posts and the escarpments together form a broken, crisscrossing pattern over the landscape. Various post-rock structures literally fit into that framework. It is not uncommon to see, for example in the pasture landscape north and south of Wilson Reservoir, a lone stone building—house, barn, school, church—on some hillslope. Or the remains of an old stone corral may tell us what history books cannot convey. Elsewhere, glimpses into the past may take the form of the remnants of an old townsite. Most of the region's old cemeteries, some enclosed with stone walls or stone posts, as at Bunker Hill, include at least a few grave markers, which together with their inscriptions help to link the region's beginnings with its development. In scattered positions throughout the area, remains of old self-sufficient homesteads may include a stone dwelling, a cave or cellar with a stone entrance and stone walls, a stone barn, two or three other stone outbuildings, a stone corral, stone clothesline poles, stone gateposts, a stone hitching post or two, flagstones for walks between buildings, stone well curbs, and stone feeding floors or hollowed-out slabs for watering or feeding farm animals. Depending on the vintage of the homestead, there might also be a stone silo, a stone watertower, or a windmill with a stone base—perhaps even a stile block (a rock platform with steps used to board buggies).

Many of the region's old farm sites and townsites are marked by a few trees, which the early occupants coaxed into growing. In fact clumps of trees beyond the streams almost certainly indicate former human habitation and probably a stone building or buildings. As Arthur Jepsen reminded us, almost every quarter section of land in many townships in the area once had at least one stone building. A few farm complexes from the past are still being maintained and occupied, some by descendants of the original owners; but it is the abandoned structures that emphasize that the old family homestead in many instances has been superseded by larger farms and fewer residences.

The role of post rock may best be appreciated by spending some time in a small area, such as a 150-square-mile area in northwestern Lincoln County and the adjacent part of Mitchell County that includes Hunter. Most of the area is drained by Spillman Creek, which empties into the Saline; but east of Hunter, along Rock Creek, the drainage is via Salt Creek into the Solomon. Several of Hunter's buildings are of post rock. Included is its largest, the two-story older part of the town's school, erected as a high school in 1918.

A livable though unoccupied L-shaped stone house built in 1874 on the Frederick Schneider homestead, about two and a half miles northeast of Hunter, belies its century of existence. B. J. Schneider, who now owns and farms the place, lives about five miles farther east, a mile east of Victor and across the road from the Lee Faulhaber homestead, which he also farms. On this homestead, along Rock Creek, the Faulhaber cave built in 1872 and the home built by Faulhaber at a later date remain in usable condition. Gone, however, is the blacksmith shop to which "the menfolks" of incoming families once found their way for feathers and wedges so much in demand or to ask Faulhaber's services in quarrying "tall posts" or building stone. Faulhaber probably made the tools and quarried the rock for the L-shaped Schneider house.

At Round Springs, about a half-mile east of his home, Faulhaber once served as postmaster. About all that remains of the town is a cemetery and one post-rock building constructed in 1914 to serve as a nondenominational church and community hall.

About three miles south of Round Springs, across the Mitchell County line into Lincoln County, the two-story house built of post rock in 1879 by James Peter Webster, grandfather of Allen Webster, stands structurally intact. No longer a residence, it is used for hay storage by Frank Webster, Allen's son, who lives across the road and farms the home place.

A mile west and a mile south of the Webster place, a stone dwelling exhibits some of the carving skills of an early-day stonemason, Emory H. Cook, Irish father of Mrs. C. Hobbs of Lincoln. Mrs. Hobbs said that her father and mother came to Kansas in 1876 and that her father began work on the house in the mid 1880s, taking several years to complete it, while "raising a family in a tent set up nearby." He built it carefully. Even though it has not been lived in for several years, no cracks mar its walls (two feet thick), inside or out. Using feathers and wedges, Cook obtained rock from a quarry southeast of his farm. He used chisel and hammer to shape the stone for the walls into pitched-face blocks, which he laid in uniform courses, using a native lime-and-clay mix for mortar. He used a tooth-face hammer to dress some of the stone used near the foundation and around doors and windows. He also did some fancy carving on stone used as trim, as shown in a photograph in the color-plate section.

It was Mrs. Melvin Schulz who called our attention to the Cook place, now owned by Tressa (Mrs. Wilbur) Wallace of Barnard. Mrs. Schulz also took us to a neighborhood cemetery to show us a tombstone on which Cook had carved the likeness of a lamb. Her interest in the stone structures of the area derives from her heritage. Her grandparents, Mr. and Mrs. Herman Will, came from Germany and "homesteaded in the area almost 100 years ago. They built stone houses and barns and fenced miles and miles of pastures and fields with stone dug from their land."

Just as the Cook cottage illustrates the artistic bent of an early craftsman, a stone wall around eighty acres of land four miles south and six west of that house represents the enterprise and ingenuity of an early settler named Bacon, who lived on that eighty acres. He was one of those fortunate enough to have a lot of timber along Spillman Creek, which ran through his property. Deciding that he would like to enclose his property within a stone wall (and perhaps collect the bounty of $2.00 for each forty rods besides) but being disinclined to do the work himself, he hit upon a trade idea. Anyone who wanted wood for fuel could chop it and haul it away "for free" if he would bring a load of stone and lay it up in the wall. That idea brought the desired results, so local folks say.

Two miles south of the Cook cottage, stone ruins record the diversity of a farm operation early in this century: upended stone slabs that once were part of two barns and a cow shed, a chicken coop, a pig shed, feeding and watering vessels for farm animals, fence posts, an implement shed. There are several flagstone walks, a stone cellar with a stone air vent, and a stone house with a 1905 date stone. Known as the McKinney place, it has been owned in succession by Louis McKinney, who bought out a homestead right in 1878; then by son William, who bought out the heirs; and now by Hazel McKinney Flaherty, daughter of William McKinney.

House of post rock on the James P. Webster homestead, known as "The Blue Stem Ranch," northwestern Lincoln County.

Chicken coop on the McKinney place, northwestern Lincoln County. Note also flagstone walk and hollowed-out stone block used as a watering vessel.

On A Journey Through the Land of the Post Rock

Four miles west of the ruins and a mile east of Ash Grove, the Verdon Peckhams live in a house built of Fencepost limestone in 1882. For many years the farm belonged to the Arthur Jepsens. In 1937 Jepsen added a rounded, enclosed, post-rock porch to the house. He used post rock for other improvements on the farm, including a round water-tower and a garage, and he quarried and set stone posts until he moved to Lincoln in 1969.

Nearby Ash Grove is now almost depopulated; nevertheless, the Methodist church, built of post rock in 1905 and in good repair, is a Sunday meeting place for area Methodists. About three miles south of Ash Grove, a lone stone building that once was a post office marks the site of Pottersburg, owned by Anita F. and Wilbur Oetting of Sylvan Grove in 1973. A half-mile east of Pottersburg, an abandoned stone schoolhouse sits in a plowed field. A mile and a half farther east and a half-mile south, a neatly landscaped farm estate is the home of the John Errebos. A huge stone barn and a two-story, box-style stone house add color and dominate the farm layout. The house was built in 1902 and 1903, the barn in 1908. The strength of both structures is in the ruggedness of their stone walls and their simple construction.

Three miles south of the Pottersburg site, on the unoccupied Meier-Zachgo estate, are several reminders of post rock's former service there: a house built in 1882, several outbuildings and building foundations, a cave with a graceful stone arch, flagstone walks, stone clothesline poles, feeding troughs, and many stone posts.

Stone house that marks the site of Pottersburg, Lincoln County, and once was that town's post office.

About three miles west of the Meier-Zachgo place, we paused to examine a stone pit silo, hoisting devices intact, at the edge of a field. About a mile and a half southwest of there, we admired what probably is the largest house of post rock in northwestern Lincoln County. It is a three-story structure of block design with wide roof overhangs. A porch extends beyond the length of one side at both ends and tends to wrap around the corners to include parts of two other sides. Once the home of Charles Shelhammer, who built it in 1913, the well-preserved house is now owned and lived in by Elmer and Evelyn Rebenstorf (brother and sister).

The Denmark community, near the southeastern limits of our 150-square-mile area, is one of Lincoln County's first permanent settlements. Denmark the town now is little more than a block of empty stone buildings and a two-story, abandoned stone schoolhouse, but the surrounding community seems prosperous enough. The little Danish Lutheran church on the rise east of town is the community's attachment to its beginning decade. The first Danish immigrants, who came in 1869 and 1870, before Indian raids could be called a thing of the past, set high priority on building a church. By 1875 some members of the congregation were quarrying and hauling stone for the purpose, donating both stone and work. In 1953, the year of the church's diamond jubilee, Ernest Andreson, at that time the lone survivor among the original workmen, provided some details for the *Diamond Jubilee* pamphlet, edited by Willard R. Garred. Andreson mentioned that he had used a big government wagon and yoke of oxen to haul stone for the foundation and that some members "handy with laying stone" had done so. The pamphlet records that "sometime during the fall or winter of 1876 Lars P. Nielsen and a Mr. Holmberg, a Swede, laid the cornerstone. . . . However, the work seemed to lag. . . . By spring only the foundation and the walls up to the bottom of the windows had been completed, and then the work ceased for a year and a half."

When work was resumed in the fall of 1878, the stone was finished by James Morgensen, A. Rasmussen, Hans Hansen, and Niels Nielsen for $25 each. Work was completed within a year, except for the windows, which were added in 1880.

Feathers and wedges were used in the quarrying, which was done on nearby property belonging to church members. The rough-hewn limestone blocks for the walls were dressed with a stone hammer, and Malcom Robinson, an early-day stonecarver of Lincoln, was engaged to do the "fancy stone cutting." The bell tower and entry on the south end of the church were not added until 1901. In harmony with the landscape and in command of its almost treeless knoll, this little yellowish-brown church has a dimension of continuity, and on a Sunday morning the bell rings out to the descendants of its founders.

The Evangelical Lutheran
Community Church of
Denmark, Kansas.

Obviously we have not mentioned all the stone structures within the
area singled out for a close look, nor have we taken special note of the stone
fence posts that line the roads. But our observations should help us to
appreciate the sentiment of Mrs. Allen Webster, who once lived in the area:
"As I view the remnants of these old houses and walls of eighty and ninety
years ago, the thought comes to my mind, those people loved this land of
America, and did the best they could with what they had."

One route to post-rock consciousness is to travel along the Solomon—
from Delphos to Glasco to Simpson and Asherville, to Beloit, and on to
Cawker City and Glen Elder and Downs—driving on a few side roads and

around a few town blocks along the way. In such a journey, the stone posts become a symbolic chain that links all post rock's uses together and to their common source: the escarpments along the Solomon. For a panoramic view, one can stand at the crest of an escarpment, for example on "Valley High Historical Point," four miles west of Delphos, the site marking one of Zebulon Pike's camp sites and vantage points some sixty years before the age of settlements and stone posts.

A landmark northeast of Delphos and east of Glasco is the Bethel Methodist Church. It was built there, two miles west of U.S. 81 and a mile south of U.S. 24, by a Methodist congregation in 1880. It is of rough-hewn Shellrock, and except for the curved trim over the door and windows it is similar in design and shape to the Danish Baptist church near Jamestown, about twenty-five miles to the northwest. The church measures thirty-six by forty-eight feet, has twenty-inch-thick walls, and can seat two hundred persons. Services were held there regularly until the 1930s. On October 12, 1970, according to the *Salina Journal* (July 31, 1971), Olive (Mrs. Jesse) Barker, now of Salina but formerly of the Bethel community, started a movement to restore the church. With donated funds and the volunteer labor of friends and descendants of former members, the church walls were tuck-pointed and other repairs were made. Mrs. Barker referred to her part in the project as "the most rewarding thing I have ever done in my life."

Entering Glasco from U.S. 24, the alert traveler will notice several modern homes of post rock from the Simpson quarries and constructed since 1950 by Joe Prickett. Downtown, older structures of post rock and Shellrock may be seen. One in particular is in good repair: the grocery store built by J. W. and Adam Studt in 1902, and until recently operated by Charles A. Studt, son of J. W., who has exchanged his groceryman role for that of a columnist with the *Glasco Sun*. St. Mary's Church, built of quarry stone in 1910, and St. Paul's Lutheran Church, built of cut stone in 1950, present contrasts in use of post rock in church construction.

Detours off U.S. 24 or Kansas 9 onto section-line roads east of Beloit almost certainly will bring into view buildings and other structures of post rock, and quite likely some will display the workmanship of Scandinavian and British immigrants who settled on the upland there. For example, in 1872 W. R. Stockard, Irish immigrant and stonemason, built a house of post rock atop a hill on his homestead, about four miles east of Beloit and a mile south of Kansas 9. That house, with additions of later dates, now belongs to O. H. Senter, Stockard's grandson. Mr. and Mrs. O. H. Senter live there.

A mile north of the Senter farm, on a hillside near Gilbert Station on Kansas 9, stands another home of post rock that has been there since the first decade of the area's settlement. It is on land now owned by Philip Doyle but first owned by E. E. Fisher, who built the house in 1878 and whose name (with the date 1879) appears on the cornerstone. Doyle, who

The W. R. Stockard house east of Beloit, Mitchell County, as it appeared many years ago; original part built in 1872 and a later addition shown. *From an old print in the possession of O. H. Senter, present occupant and grandson of Stockard.*

has lived in the house all his life, said that his parents bought the farm from Joe Gilbert, a member of the Gilbert family who bought the place from Fisher and who built a grain elevator (Gilbert Station) on the branch of the Missouri Pacific Railroad that runs through the property. "The old house was a favorite stopping place for families coming west from back east, having one of the few never-failing water wells in the county," Doyle pointed out. The stone house, a stone barn (built in 1879), and other stone buildings on this neatly landscaped farm are in excellent condition.

Beloit, county-seat town at the intersection of highways U.S. 24 and Kansas 14 and 9, is the hub of the Solomon Valley. It has retained its post-rock image to a marked degree. That Beloit literally "was built of stone," as some old accounts would lead us to believe, becomes more credible to us as we see in block after block Fencepost limestone in use—if not for entire buildings, then for foundations, retaining walls, walks, porches, steps, trim, or decorative objects. In the business district, stone buildings from several past periods are in evidence. Bernice (Mrs. Myron) Chapman told us that the members of her family recently toured the back alleys of the town's business district and confirmed what they had suspected: that most of the buildings have their original foundations and walls, that "only the fronts on most of the buildings have been changed or modernized. We saw almost no change in at least two buildings—the old Bunch Drug Building on the corner of Main and Mill and the Holloway Wholesale Grocery Building on the south end of Mill Street."

Early buildings at the Girls' Industrial School, established in the late 1880s at the north end of town, were of native limestone. Of those original buildings, Denis J. Shumate, superintendent, said only one remains: West Cottage, remodeled in 1953 and now called Shadyside Cottage.

Two of Beloit's earliest church buildings still accommodate congregations: the Christian Science church, built as the Methodist church in 1874, and the Baptist church built about the same time (a newspaper account says 1872; Andreas's *History of Kansas* says 1875). Both are of post rock. Of the town's occupied and maintained old stone houses, none is thought to be older than the semi-mansion on Kansas 14, at the turn on Eighth Street. Mrs. Margaret Hanni and daughter Miss Jeanne Hanni, present occupants, said that the original part, one room upstairs and one room down, was erected in 1873 by C. A. Perdue on land that had been granted to Charles H. Morrill the previous year. Perdue purchased the house and property in 1879 and added eight rooms (four upstairs and four down). Since 1955, when she bought it, Mrs. Hanni has restored this house of post rock, as nearly as possible, to its former majesty.

Beloit's most celebrated house of the past, the elegant two-story structure opposite the courthouse, was built between 1879 and 1884 for the town's first banker, F. H. Hart. Constructed of Fencepost limestone with Shellrock trim, the Hart place, now owned by A. L. Street, is a showplace from the Victorian era. Indeed it is now in the National Register of Historic Places. The rough-hewn post-rock blocks in its walls are pitched face and fit neatly into courses. Smooth, grayish-white Shellrock blocks arranged along the sides of the house's forty-nine windows and in corner quoins contrast with the buff Fencepost, giving a checkerboard effect. The house has decorative roof overhangs, a two-story bay window under a gable roof, and several porches with lacelike arch openings. The house cost less than $12,000 to build, though landscaping and other finishing details added $2,000. James Kinneard did the stonework.

Beloit's residents like to point out that the massive, rugged Mitchell County Courthouse truly represents its governmental scope, in that the Fencepost limestone for its walls came not from one quarry but was hauled by the wagonloads from "all over the county." Of Richardsonian Romanesque design, it was built in 1901 of "eight-inch pitch-faced stone," as noted by Fred Wessling. J. C. Holland was the architect. F. A. Cooper, a Kansas cartoonist, in his "It Happened in Kansas," widely published in Kansas newspapers, said of its clock tower (Cartoon 1107, 1958): "The stone tower on the Mitchell County Courthouse in Beloit is thought to be the tallest native stone structure in the state. It houses Beloit's 'Big Ben,' which gives the time of day on all four sides of the lofty structure."

Another massive edifice is St. John's Catholic Church, erected of pitched-face post rock between 1901 and 1904. Wessling called our attention

The Hart House, owned by A. L. Street, Beloit. *Photograph courtesy of Alan Houghton.*

St. John's Church, Beloit. *Photograph courtesy of Alan Houghton.*

to the flying buttresses that support the stone arches. Like the celebrated St. Fidelis Church in Victoria, it has twin towers (110 feet tall, compared with the 141-foot towers of the Victoria church).

Modern use of Fencepost limestone in the Guaranty State Bank, the Presbyterian church, and several homes carries on the Beloit tradition.

When Beloit was having its post-rock building boom, so were Cawker City and Glen Elder. Glen Elder, at the eastern end of Waconda Lake (Glen Elder Reservoir), has its row of limestone buildings in "Nash Block, 1908," as well as others of that vintage elsewhere around the town square. The service station built in 1926 by the town's pioneer in the automotive industry, E. W. Norris, represents a departure from the architectural style of the older buildings. Castlelike in appearance, it is in fact unusual. "E. W. Norris, 1926" is inscribed in both ends of an open-ended archway that has the appearance of a drawbridge joining the major part of the station to two watchtowers. Townspeople say Norris got the idea for his miniature medieval castle while serving with the American Armed Forces in Europe in World War I. In Luxembourg he saw a castle he liked. He studied it, and when he returned to his automobile business in Glen Elder, he drew up plans for a service station that would resemble that castle as much as possible. The gas pumps under the "drawbridge" are gone now, and the "castle," still projecting a medieval image, is the office of an insurance company.

Near the town entrance off U.S. 24, an abandoned building (once a service station, later a doughnut shop) has walls of flagstone slabs in "rubble" construction: fitted together, their bedding planes facing outward in no particular pattern. This style was common in the 1940s, according to Wayne Barnett.

One of Cawker City's quaint buildings is the old public library built in 1884 for the Ladies Hesperian Club and now in service as a local museum. A small cottage built of post rock, at the end of a curved flagstone walk lined with shrubs, it has an old-fashioned charm. It has been entered in the National Register of Historic Places.

In Downs, at the west end of Waconda Lake in Osborne County, business "goes on as usual" in a few post-rock buildings from the pre–World War I era. The same can be said for Osborne, about twelve miles southwest of Downs. Osborne, slightly west of Fencepost limestone's outcrop area, has at its perimeter outcrops of Fort Hays chalk, also used in many of the town's buildings. One of the most impressive and colorful of Osborne's post-rock buildings is the Osborne County Courthouse, built in 1907. Richardsonian Romanesque in design, it is dominated by a clock tower typical of that period. The brown streaks in the pitched-face blocks of its walls contrast distinctly with the light buff of the rest of the limestone. Post rock used in the building was shipped in from a neighboring county.

Glen Elder's "castle" service station, 1926. Photographs show (top to bottom) yard where stone was cut—note mode of transporting blocks and also public school building of post rock (no longer standing); under construction; and open for business. *Photographs courtesy of Wayne Barnett.*

Cawker City's public library for more than eight decades. Now a museum; listed in the National Register of Historic Places. Architectural style: Kansas vernacular.

To enjoy a post-rock motif generally similar to but specifically different from that of Solomon Valley, we may start in eastern Lincoln County, perhaps as far east as Beverly or Shady Bend, and drive westward on Kansas 18 through Lincoln, Vesper, Sylvan Grove, Lucas, Luray, and Waldo to Paradise—and a few miles beyond. Before entering Lincoln, we might stop at the roadside park to observe the unusual fence around the adjacent cemetery: carved limestone posts with drilled holes through which iron pipes extend laterally. Some of the old tombstones in the cemetery are of native rock carved by Malcolm Robinson, the talented early-day stonecutter who did the "fancy stonecutting" for the Danish Lutheran church at nearby Denmark.

A drive through Lincoln's business district and around several blocks shows that Lincoln on the Saline, like Beloit on the Solomon, relied heavily on post rock from the town's beginning to recent times. Elizabeth N. Barr, in *Souvenir History of Lincoln County, Kansas,* 1908, did not exaggerate: "Nearly all the business houses in Lincoln are built of native rock [Fencepost limestone], and except for the few frame structures brought over from Abram, they have always been."

The year 1886 was one of the most important in Lincoln's history. In that year the Union Pacific reached town, Kansas Christian College opened its doors in a two-story stone structure erected the previous year, and 125 buildings were erected. In her *Souvenir* booklet Barr noted that the pop-

One of Lincoln's business buildings of post rock; built as a bank the year the railroad reached the town. Various architectural forms incorporated in design.

Lincoln County Courthouse, Lincoln; in use since completed, about 1900.

ulation doubled and the wealth trebled. Several of those buildings (the two-story college building is not among them) are extant and in use. One of the most prominent bears the inscription "Lincoln State Bank: 1886"; it is now an abstract office. The building was the work of Malcolm Robinson. In its walls near the entrance are two polished slabs of stone locally called "Lincoln marble" but described by Harvey Roush as "a dense, nodular limestone appearing above the Fencepost bed in the outcrop." Robinson must have had quite a business based on this nodular limestone. In the *Lincoln Beacon* of March 3, 1887, he is referred to as "Proprietor of the Lincoln Marble Yards, has quarried and polished a large quantity of this marble, working it into tombstones, mantels, tabletops, etc."

One of the busiest places in Lincoln for many years was the three-story stone mill erected in the 1870s on the Saline west of town. According to the *Souvenir* booklet, the mill, still operating in 1908, was "one of the most beautiful spots in town." Torn down some years ago, the mill was representative of the many water-powered mills (generally built or partially built of stone) that once existed at almost every stream crossing. Grist mills and flour mills were considered essential to the area's survival.

Buildings of post rock erected in downtown Lincoln between the 1890s and World War I include the city hall (1913), the public library (1913), and the Lincoln County Courthouse (1899–1900). Like the courthouses in Beloit and Osborne, the one in Lincoln has a clock tower. The tower, however, almost an angular cupola, is located centrally, and the building differs slightly from the other courthouses of the area. Two curved walls on each of two sides extend above the main roof, each becoming a completed cylinder with a conical roof.

The stone building housing the *Lincoln Sentinel-Republican* was erected as a plumbing shop in 1914 but has been a newspaper office since 1922. It acquired a fresh appearance in 1963, when it was sandblasted. To bring post rock up to date in Lincoln, we must add the Post Rock Motel, built north of town in 1956.

Along Kansas 18 west of Lincoln it is possible to find these holdovers, among others, from Lincoln County's past: a small stone building with a dirt roof supported by thatched twigs—obviously once used as a blacksmith shop; a two-story stone house with eight gables built in 1896 and standing with dignity but alone in a plowed field on the "old Watson place" (now owned by Herman Crawford) south of Goldenrod; the tiny Simmons cemetery (about six miles west of Lincoln, three-tenths mile north), marked by a few stone posts and a few tombstones, including one for a soldier who had been a drummer boy in the War of 1812; the Vesper school building, erected in 1914; the Vesper cemetery, enclosed by pillar-type posts of post rock; and wall remains of a stone ranch house (on the "A. S. Sutton" place, about four miles east of Sylvan Grove) built before 1886, the year the rail-

Along Kansas 18 west of Lincoln: street scene in Sylvan Grove, 1952 (upper photograph); Gorge school, west of Sylvan Grove, 1973.

road reached Lincoln. The walls, two feet thick, show in cross section a strength that seems almost defiant (observe photograph in color-plate section); after all, they are the remains of a house that did not crumble but was "quarried" for building block instead. To view the artistry of those remains is to respect the craftsmanship of the stonemason, whoever he was, who built the house. Frank B. and Albert Walker, most recent owners, are zealously guarding their possession.

In Sylvan Grove stone buildings on the main street have a turn-of-the-century look. The original townsite, however, was southwest on the Saline, on what is now the Robert Wenthe property. Some of the region's oldest standing stone posts—posts with holes for wood plugs, with plug remnants in a few of the holes—are curiosities along the route between the old and present town sites.

A two-story stone store built in the 1870s at the old site now is used as a farm outbuilding. Before the railroad reached this part of Lincoln County, Germans, Bohemians, and people of other nationalities came from thirty miles distant to shop at this store, the only one in the area, and to have their grain ground at the mill that once was located nearby.

Water tower of post rock, near entrance to Paradise.

Side trips north and south of Kansas 18 in the vicinities of Lucas, Luray, and Waldo are especially recommended for examples of stone posts set and stone buildings erected in the past century. The Ralph Goodhearts, for example, once operated a farm in the Wilson Reservoir area, where they lived in a limestone house built in 1911. When the reservoir was constructed, they moved into another limestone house (on a farm west of Lucas), part of which was built about 1880 and part about 1900. Mrs. Goodheart notes that the home is "all in good condition" and that on the property are "stone posts 8 in. by 12 in. by 10 ft. quarried about 1880 and also a stone quarry, which is no longer worked."

Drives south to Wilson Reservoir are full of opportunities to see post-rock country in its broadest spectrum: land forms and outcrops where post rock can be observed in relation to the other beds of Greenhorn limestone; stone posts above the exposed ledge of post rock; as well as stone buildings and remains of stone buildings. There are stone-arch bridges along Paradise Creek and its tributaries. Finally, at the end of the Saline route, there is the stone water-tower that marks the entrance to Paradise.

Except for Ellsworth, most towns in the stretch of post-rock country between the Saline and the Smoky Hill are on the divide that separates the streams' drainage basins rather than in their valleys. Because Interstate 70 follows this divide across post-rock country, Wilson, Dorrance, Bunker Hill, Russell, Gorham, Walker, Victoria, and Hays might be called the area's superhighway stops.

Ellsworth, the exception, is in the Smoky Hill Valley and belongs perhaps more to Dakota country than to post-rock country. Still, it is a transition town, where brown sandstones and buff limestones blend. For example, the Ellsworth County Historical Society buildings—the old Dakota sandstone house built in 1875 and livery stable built a few years later—on old Front Street have some post rock in their walls, and there are flagstone walks of post rock or "flagstone limestone" on the museum grounds. Ellsworth's beautiful little Episcopal church built of Dakota sandstone in 1897 has post-rock trim.

One of Ellsworth's post-rock homes is particularly famous: the old "manor house," once the self-sufficient homestead of the Arthur Larkin family. On a knoll southwest of town, the house was built for Larkin in the mid 1880s; he owned the renowned Ellsworth Grand Central Hotel. Larkin's grandson, Arthur Larkin, now a New York advertising executive, in an article written for the *Kansas City Star*, May 27, 1965, reminisces about his boyhood at this place. He remembers a huge barn of Dakota sandstone, destroyed by a tornado. He remembers there were gas wells, which provided fuel for heat and light. There was a water system, a smoke house for curing hams, and a multiroomed basement for storing vegetables and fruits. He remembers the manor's

> thick limestone walls and wonderful big, light-bright rooms . . . its big 2-story center hall, elaborate ceiling molding, six marble fireplaces and exterior carvings. . . . built during the post–Civil War boom in a style reminiscent of Italian villa architecture of the same period. . . . its only concessions to the ornate Victorian architecture of the time were the stone carvings and the tower which rose four stories, providing a view that included two towns and miles of fields and woods.

That house, now occupied and owned by Loren Dees, operator of the Garden Motel, is being restored.

The original part of the Adolph Vopat residence three miles east of Wilson, in northwestern Ellsworth County, has the date 1882 chiseled in a stone over a window. Jakub Vopat from Bohemia built that part of field stone he picked up on the homestead. Later, quarried post rock was used in an addition, and an inscription on a smooth stone high on one side of that addition clearly reads: Barbora Vopat, 1900. Still later a frame section was added. Lila (Mrs. Adolph) Vopat told us that four

The Arthur Larkin house of post rock, near Ellsworth, as it appeared soon after it was built in the mid-1880s. *Photograph courtesy of George Jelinek.*

generations of the Vopat family have lived in the house, recently remodeled so that the stonework is exposed on the interior of several walls. She said that when they removed the plastering they found the two-foot walls solidly in place. All they had to do was to remove the old lime mortar and replace it with cement.

Wilson, once conveniently near post-rock quarries on the north and on the south, has some handsome exhibits of post-rock business buildings on both sides of the Union Pacific Railroad tracks, which bisect the town. The Wilson railway station of dark-brown Dakota sandstone offers a contrast in stone architecture. (The sandstone blocks in the station first were used in the hospital at old Fort Harker, near Ellsworth.)

That times were getting better in the mid-to-late 1880s in this Bohemian center—today's Czech capital of Kansas—can be surmised from

The Vopat residence east of Wilson, showing the original part, built of field stone in 1882.

Midland Hotel, Wilson, Kansas, in 1904. An example of Renaissance architecture in Kansas. *Photograph from old print lent by Mrs. Agnes Hill, owner-operator.*

the substantial stone structures that are holdovers from that decade. At least one holdover, a two-story post-rock building north of the tracks, has the distinction of still being used for its original purpose under its original name. "Wilson State Bank, 1886" appears on two sides at the top of the well-kept structure, which has only hints of ornateness.

One block on old U.S. 40, south of the tracks, consists essentially of limestone buildings dating from the turn of the century. One is the old opera house. The corner building, with its series of closed archways two stories high, was once a bank building and is now a recreation center. The date stone reads "Weber & Pierano, 1904." A round building of post rock in one of the downtown alleyways was a reservoir for the town's water supply, later a city jail.

Wilson's colorful but unpretentious Midland Hotel has been in business since its sign was hung out at the beginning of this century. It is the reconstructed Powers Hotel, built of pitched-face post rock from a quarry south of town between 1894 and 1898. The interior, but only the interior, was destroyed by fire in 1902. Its walls are the walls of the Midland Hotel. The reddened brown streaks around the window openings are the only visible effects of the fire. Samuel Anspaugh, a stonemason, did the reconstruction in 1902 when he was sixty-five; then he and his wife lived there but leased it to an operator. Later it was owned and leased by a group of Wilson businessmen. Mr. and Mrs. John Hill, who have been operating it since they bought it in 1960, are the only owners who also have operated the hotel.[3] Mrs. Hill explained her attachment: "We love this place, and we are making an effort to preserve it. Now we wouldn't live anywhere else. We want to prepare a brochure on it, and we might include a few recipes of Czech dishes. Occasionally we serve Czech meals."

Side trips north and south of Interstate 70 between Wilson and Russell —particularly north of Dorrance and Bunker Hill and along the South Shore drive of Wilson Reservoir—invariably lead to exposures of Greenhorn strata that pronouncedly show the Fencepost ledge. Anyone geologically inclined should enjoy stops to examine some of those exposures, for example some along the paved road leading north from Bunker Hill. Persons looking for them can recognize hillside indentations, scars if you please, that offer proof the area once was pock-marked with quarries. Traces of former local uses of the quarried rock range from stone posts to ranch houses and barns on the rural scene and include flagstone walks, residences, and business buildings in Dorrance and Bunker Hill.

In Bunker Hill old stone business buildings still in use include the post office, built as a bank in 1886; the Masonic Hall, built about 1890 as a general hardware store; Neill's Cafe, built as a drugstore in 1916; and Bunker Hill Grocery, built as the Eyler-Gross General Merchandise Store in 1887. One of the town's best-preserved structures is the building of

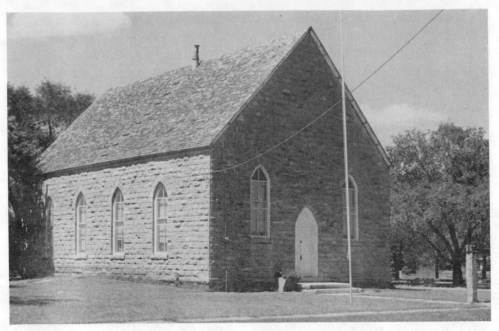

Bunkerhill Museum and sign post of post rock.

sledged post rock that was the Lutheran church from 1880 to modern times and that now houses the historical museum. There has been no major construction with stone in Bunker Hill since the school auditorium-gymnasium was built, as a WPA project, in 1939.

Evidently soon after members of the Northwestern, or Ripon, Colony got off the train at Fossil Station (now Russell), about eight miles west of Bunker Hill, in 1871, they went to the outcrops for building stone. That same year Benjamin Pratt, the colony's president, wrote to the *Free Press* of Ripon, Wisconsin (the group's former home), according to the *Russell Daily News* June 1, 1971, "Stone is very abundant and of the best quality and can be delivered in town for $3 a cord."

Among Russell's houses built of post rock or of stone brick (sawed Fairport chalk) within that first decade and extant in the 1970s, at least two have been referred to as showplaces: the Copeland-Ackerman house at 324 East Fifth, built in 1878; and the house at 11 Front Street, built in 1879–1880 by the town's first storekeeper, H. W. Tusten. The Copeland-Ackerman house, also known as the M. P. Miller house, now belongs to Helen M. Holland and is the residence of Mr. and Mrs. Bert Hitchcock. Kate Lee Ewing, Russell historian, noted that C. C. Hulet, Russell's first sheriff, designed the house. The Tusten house, now painted white, is the property of its occupants, the Loren Furneys. Gertrude (Mrs. Loren) Furney, who is an artist, pointed out that another artist, wood engraver

Hubert Deines, who has received many national honors, lived in this stately stone house during his growing years.

The nucleus of Russell's present First Congregational Church, at Sixth and Kansas streets, was erected in 1878. The modern part, of brick, blends into a lower wall of rough-hewn native limestone. The basement is known as Ripon Hall, not only honoring the original colony but also reminding today's congregation this church was the first one organized in Russell.

Like the inner-region county-seat towns of Lincoln and Beloit, the county-seat town of Russell through the years has maintained its allegiance to post rock. Even the laws of town and county are administered from post-rock buildings. The city building, of sawed stone, was erected in 1941 (at a cost of $54,673, according to Robert E. Meneley, city clerk). The Russell County Courthouse, as previously noted, was built, using quarry stone, in 1901 and modernized, using cut stone, in 1948. About a block east of the courthouse Russell youngsters attend classes in functional Ruppenthal Middle School, of sawed and pitched-face post rock, erected as a high school in the late 1930s. East of the courthouse the castlelike structure of pitched-face post rock, with stone posts all around, was built in 1904–1905 as the county jail and sheriff's residence. Now the Fossil Station Historical Museum, it is the place to get information on where to go to observe post-rock geology and post-rock products in Russell and environs. An unusual exhibit is the post-rock stepping stone with hitching ring in the yard of Georgiana (Mrs. Frank) Bell at Elm and Wisconsin streets; it remains where it was placed on the curb (when the street was wider) in the horse-and-buggy days.

Regardless of what their fronts suggest, Russell's business buildings, as seen from the alleys, tend to be a series of walls of native limestone. The stone posts, positioned on downtown streets to commemorate the town's hundredth birthday, thus are a modern front to a backdrop of limestone buildings.

Ruppenthal Middle School, Russell. Constructed of Fencepost limestone as a high school in the late 1930s, it is a good example of prairie-style architecture. *Photograph (by Paramount Studio, Russell) courtesy of* Russell Daily News.

Stepping stone in yard of Mrs. Frank Bell at Elm and Wisconsin streets, Russell; once used by members of the G. W. Holland family to board their buggy.

Fossil Station Museum, Russell, maintained by the Russell County Historical Society. Kansas vernacular in design.

Hays, post-rock country's western portal on Interstate 70, provides an opportunity to study a tradition in use of native limestones on a frontier outpost and on the campus of the region's public institution of higher education, Fort Hays Kansas State College. We might start with a leisurely stroll across the campus, on part of the grounds of old Fort Hays Military Reservation. We observe that Fencepost limestone is the principal building stone, though Fort Hays chalk from west of Hays and limestone from eastern Kansas have been used also. In 1903, a year after the Western Branch of the (Emporia) State Normal School (the first official title) was created, Fencepost limestone was picked and hauled to the campus to erect Picken

Picken Hall, administration building at Fort Hays Kansas State College, Hays; erected of post rock in 1903.

Hall, the administration building. The walls of Picken and two other post-rock buildings (Martin Allen and Rarick halls) erected before World War I are of straight quarry stone (untrimmed); they are laid up in such a way as to show drill marks and bedding planes but not the fossils they contain. Picken Hall, basically blocky and simple in design, nonetheless has a portico supported by Ionic columns. Sawed, chipped-face or split-face veneer blocks in the walls of the Fencepost buildings erected since 1950—Davis Hall, Memorial Union, the new part of Albertson Hall, Wooster Place Apartments, Malloy Hall, Gross Field House—are set so that the split center of the rock layer is exposed; revealed are many fossils.

Over a number of years, students who enrolled in geology courses taught by Professor Myrl V. Walker saw the stone walls as a part of, as well as housing for, the learning process. Professor Emeritus Walker, long-time director of the college's Sternberg Memorial Museum and formerly head of the geology department, gave his students a few lectures on historical geology and then sent them on a "wall search" to find clues as to the limestone's makeup and origin. When they could find and identify the clam *Inoceramus labiatus*, coiled snail-like ammonites, shark teeth, bone, and a few other fossils exposed in the post-rock walls, he took them on a field trip to look at the rock in its natural outcrop.

Fossil clam *(Inoceramus)* in section of limestone wall of Gross Field House, Fort Hays Kansas State College.

Just as the buildings on the college campus represent a blending of the moderately old with the new, as well as continuity in the use of native limestone, so those in Frontier Historical Park (south across U.S. Alternate 183) represent a contrast between the very old and the new. The Fort Hays blockhouse, built of Fort Hays chalk in 1867, has as its modern companions a visitors' center, built of Fencepost in 1970, and the "Monarch of the Plains" statue, with its base of Fencepost, erected in 1967.

Driving through Hays, one can recognize various public buildings and residences, some old and some new, of native limestone. Some are of soft Fort Hays chalk, from the bluffs west and north of town, but more are of the harder Fencepost from outcrops to the east and south. The traveler inquiring about structures of Fencepost and associated limestones in the Hays area, incidentally, likely will be told they are of "Benton limestone," a term once widely used.[4] Some structures in Hays, as in other towns, are of stone salvaged from razed buildings. Included in that category is one section of the stone part of the First Presbyterian Church, erected in 1878–1879 and added to in 1913. The Rev. Blaine Burkey, historian for the Ellis County Historical Society, informed us that the source of the rock for the 1913 addition was Victoria's Episcopal church built in the 1870s by George Grant. The stone part of the church, of Fort Hays chalk and scattered Fencepost limestone, was entered in the National Register of Historic Places in 1973.

George Grant Villa, south of Victoria, 1973. Clothesline pole in foreground. Charles L. Hall, architect, describes the villa as an "English country house in the Kansas vernacular."

When we pause in front of the gold-domed St. Joseph's Catholic Church, a composite of Gothic and Romanesque architecture, of Fort Hays chalk on a post-rock foundation, we may wish to turn our attention to the plains "cathedrals," retracing our route east to Victoria.

Approaching Victoria from the north, we drive past St. Fidelis Cemetery into the part of town that was once Herzog, and there in full view is the gigantic "Cathedral of the Plains." We may return for a tour by the Rev. Jordan S. Hammel, retired pastor and tourist guide. But for now let us pause only briefly and then drive on through the south part, the original Victoria, and follow the paved road toward Pfeifer until we cross Big Creek. There we turn left, and in approximately two miles reach George Grant's villa. The Scottish nobleman and his dream died here in 1878, about five years after the villa was completed. Still, Grant is remembered for introducing, at this place, Aberdeen Angus cattle to America, and a monument near his grave at the south edge of Victoria lauds him for that. The villa and farm buildings, constructed of post rock from the valley of Big Creek, are so well preserved that few would guess their age. The Baier family surely can be given much credit for that: first, Mr. and Mrs. Moritz Baier, who bought the place in 1897 from Grant's niece, Margaret Grant Duncan, and lived there until 1935; then their son William and family; and then grandson Paul and family since 1972, when Mr. and Mrs. William Baier retired and moved to Victoria. But although the villa survived, it is part of a venture backed by wealth that for its planner did not succeed.

Looking north to Victoria at the steeples of St. Fidelis, we are reminded that that edifice, also of Fencepost limestone from Big Creek valley, appeared only after the poor Volga Germans had struggled more than thirty years to attain some wealth.

Both the villa and the "cathedral" are in the National Register of Historic Places. Most of the one hundred sections of land selected for an "England in America," however, have become a part of the German success story. In fact within a fifteen-mile radius of the villa, ten "cathedrals," all of post rock or predominantly of post rock and most replacing earlier stone churches, were built before the end of World War I (see accompanying table).

CATHOLIC CHURCHES OF STONE, NOW STANDING, WITHIN A FIFTEEN-MILE RADIUS OF GEORGE GRANT'S VILLA SOUTHEAST OF VICTORIA

Town	Church	Date dedicated
Munjor	St. Francis	May 25, 1890
Catherine	St. Catherine	October 6, 1892
Gorham	St. Mary's	December 25, 1898
Schoenchen	St. Anthony	June 13, 1901
Emmeram	Sacred Heart	December 1901 (closed August 6, 1967)
Hays	St. Joseph's	June 13, 1904
Liebenthal	St. Joseph's	May 28, 1905
Walker	St. Anne	November 30, 1905
Victoria	St. Fidelis	August 27, 1911
Pfeifer	Holy Cross	May 3, 1918

Note: The stone Catholic church at Ellis—St. Mary's, built in 1913—is not included in this list because Ellis is not in post-rock country but in the Fort Hays chalk area west of Hays. Also, churches now in service at Loretto (St. Mary's built in 1928) and at Antonino (St. Mary's, built in 1952) are of brick.

Source: The Rev. Raphael Engel, Capuchin, Hays.

Lining the roads leading to those parish churches are many stone posts, stone buildings, and other evidences of post rock's role in improving the economic status of the Volga Germans.

A brochure prepared by civic groups of Victoria describes the "Cathedral of the Plains," the largest of the churches:

> The massive Romanesque structure stands in the form of a cross, facing west, its majestic towers dominating the prairies for miles around. On the facade above the rose window a stone statue of St. Fidelis keeps watch. The "Cathedral" is 220 feet long, 110 feet wide at the transepts and 75 feet at the nave. Its ceiling is 44 feet above the ground, and the towers rise 141 feet. The seating capacity of 1,100 made it, at the time of its building, the largest church west of the Mississippi. From an artistic point of view the outstanding feature of the church is the blending of classical lines of traditional Christian architecture with the simplicity of the pioneer west.

The impressive St. Fidelis is the third stone church to serve the Volga Germans who in 1876 founded Herzog. Their first stone church, which replaced a frame lean-to, was 60 by 30 by 16 feet. It was built with major assistance from the Hon. Walter C. Maxwell, a Catholic Englishman from the Victoria colony. When that church proved too small for the congregation, it was replaced by another, 168 by 46 by 35 feet, built in 1884 on ten acres of land donated by the Kansas Pacific Railway Company. By 1900 the congregation was outgrowing that church, which could seat six hundred. The need for a larger church at that time could be attributed in part to the presence of a monastery established in 1878 by the Capuchin-Franciscans from Pennsylvania. The Herzog (Victoria) complex thus became a mother parish with influence over a wide area.

Herman Linenberger of St. John Rest Home, Victoria, reportedly the only person still living in 1973 who had worked on the "Cathedral of the Plains," was employed as a carpenter the entire time the church was being built, from 1908 to 1911. We obtained from him some information on construction details, and he checked the accuracy of information from other sources. He verified, for example, that each male parishoner over twelve years old was assessed $45 and six loads of stone, some large families hauling as many as seventy or eighty loads; that the country's foremost church architect, John T. Comes of Pittsburgh, Pennsylvania, drew the church plans, which were then modified (for the western touch) by the state architect, John Marshall of Topeka; that 125,000 cubic feet of native rock (post rock) was hauled primarily from a quarry on Big Creek south of Victoria (near the Grant villa); that some stone was imported, including Indiana limestone for ornamental work and sixteen Vermont granite pillars; that except for one stonecutter all quarrymen, stonemasons, carpenters, and other laborers were hired locally, mostly from within the parish; that all

Saint Fidelis Church, or "Cathedral of the Plains," Victoria, Kansas.

work was done by hand except that a gasoline motor was used to run a fan saw to work on wood for the ceiling; and that the total cost, including construction and interior furnishings, was $225,000 (some sources quote $80,000 or $132,000 as actual money outlay).

Linenberger said that quarrymen, using hand-powered braces and bits and feathers and wedges, split out slabs of post rock averaging two to six feet long (some perhaps as long as nine feet) and eight or nine inches wide, the thickness of the rock layer. About eight slabs, two layers of four slabs arranged side by side, were placed on a wagon bed and hauled to the building site to be dressed by about six stonemasons and two or three dozen cutters. After the slabs were cut into various lengths—from eighteen to thirty inches—a worker would smooth two sides and the ends of each block, then use a chisel to make it the proper size, chipping one face so that when all blocks were fitted into the wall the building would have a pitched-face appearance. It took one man from forty to sixty minutes to finish one block, which would weigh from fifty to a hundred pounds.

Explaining that the walls of the church were of double thickness, about eighteen inches total, Linenberger noted that stone from the previous church, which had been on the same site, was used for the inner wall. Ramps, scaffolds, and a block-and-tackle system were used to lay up the walls. When the walls were five or six feet high, two hoists were used, one being loaded as the other was being unloaded at the stone-laying level. Horse power was used to raise and lower the hoists. Mortar was a mix of purchased cement, about 4,000 loads of sand obtained locally, and the "hard" waste from the stone-cutting operations.

For some of his carpenter work (the church contains 150,000 linear feet of lumber), Linenberger received as much as $1.35 an hour. He said that quarrymen and other laborers generally received less.

In this "cathedral" area between the Saline and the Smoky Hill, most parochial schools, rectories or parish houses, and other church-connected buildings, including a monastery at the Victoria site, also were constructed of post rock. The Volga Germans, who had a penchant for things of permanence, replaced their sod houses with stone as soon as they were able to do so. Some became adept at special uses, such as making ornamental pieces or carving. Herman Linenberger's brother John was endowed with carving ability. In that capacity he assisted the stonecutter brought in from outside to do ornamental work on St. Fidelis, and examples of his work may be seen in buildings in the vicinity of Victoria and on tombstones (including that of his father, Joseph Linenberger) in St. Fidelis Cemetery. On a wall of the monastery built on the church grounds in 1902 is a bas-relief with a two-toned effect, which John Linenberger achieved by splitting a slab of Fencepost limestone through its brown streak and smoothing it so that when he carved into it, the light buff under the brown

Bas-relief in wall of monastery adjacent to St. Fidelis Church. Carved by John Linenberger in post rock so that the raised part would appear brown and the recessed part, light buff.

would show. He thus obtained the reverse effect from that achieved by Felten in the bas-relief on the bison pedestal at Hays.

What post rock did for the Volga Germans in the St. Fidelis parish, it did in various degrees for those in the surrounding area, south to Pfeifer, Liebenthal, and Schoenchen, north to Catherine and Emmeram, west to Munjor and Hays, and east to Walker and Gorham. The Rev. Emmeram Kausler, who was priest in several of the parishes at various times in the late nineteenth and early twentieth centuries, was also an accomplished stonemason and, as stated by the Rev. Raphael Engel, "did his own blueprinting and actually supervised the building operations of a number of these churches." St. Anthony Church at Schoenchen is a good example of post-rock construction as conceived by this stonemason priest.

Before leaving the area, we single out one more example of post rock's use in a towering "cathedral": Pfeifer's Holy Cross Church, built between 1916 and 1918 to replace a smaller stone church. It is an outstanding Gothic structure in the Smoky Hill Valley, and its main steeple (one of three) rises 165 feet above ground level. Under its immediate protection are a two-and-a-half-story school and a rectory of post rock, and visible within its orbit are thousands of stone posts and other post-rock structures, which seem to be paying homage to the multisteepled church.

Regarding the Pfeifer settlement, we quote Father John M. Poell, who in a church history, *The Cross in the Valley*, records that as the immigrants moved toward the valley, "in the distance they could see a range of hills rolling out and away from nowhere. One of the youngsters fell over a huge outcrop of stone. . . . The men examined the outcrop and found it to be excellent material for their future homes."

Father Poell mentions post rock often in his historical account, noting its extensive use as building material, the volunteer stonework done on

Saint Anthony Church, Schoenchen. Designed by and built under the supervision of the Rev. Emmeram Kausler when he was priest there, about 1900.

the first church, and the many miles of stone posts that meant much to the settlement's welfare. In paying respect to post rock as the "people's resource," he gives us this perspective: "These good people made good use of nature's building materials. Trees were far and few. So, they split boulders [slabs] of this stone for fence posts. This is a picturesque sight in Ellis County. You see these heavy-bodied posts in every shape and weathered color standing like lonely sentinels [holding up barbed wires] on perpetual guard."

Overlooking the Smoky Hill Valley from north of Pfeifer; tall spire of Holy Cross Church pinpoints Pfeifer.

BELOW THE SMOKY HILL, ONTO THE REACHES OF THE ARKANSAS

The southwestern part of post-rock country is in the drainage area of the Arkansas River. It is defined on the north by the divide that curves northwest from Lorraine (Ellsworth County) to Beaver (Barton County), thence southwest toward Olmitz and through Rush County north of La Crosse into northern Ness County. The southern limits of the area extend from Holyrood (Ellsworth County) through Hoisington (Barton County), southwest through Rozel (Pawnee County), into northern Ford County near the entrance of U.S. 283 into that county. The western boundary zigzags north across Hodgeman County several miles west of Jetmore, then extends northeast between Ness City and Bazine to the vicinity of McCracken.

LAND OF THE POST ROCK

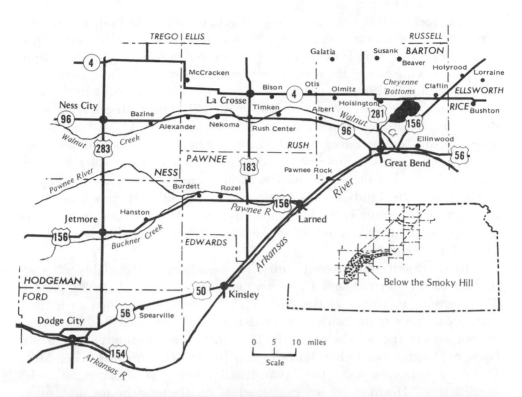

Barn of post rock, in northeastern Barton County at the Ellsworth County line, built by John Altenbaumer near the turn of the century. On property owned by Mrs. Mary Zajic and son Derald.

Post rock in this part of the region generally is pale buff, the streaks very pale brown and mostly off center; commonly there are several brown streaks. Post rock here is not readily distinguishable from Shellrock. There are evidences aplenty of the former economic importance of both limestones in fencing and building projects. St. Catherine Church, built in 1901 in the Dubuque community at the Barton-Russell county line northeast of Beaver, is one of the area's elegant edifices of post rock; it has been nominated to the National Register of Historic Places. A well-kept post-rock business building is the one occupied by the telephone company in Hoisington. As of 1973 that building was about ninety years old, according to resident Warren Tindall. An example of an excellently preserved rural structure of limestone is the barn built by John Altenbaumer on his farm in northeastern Barton County, at the Ellsworth County line, around the turn of the century.

In northwestern Barton County, most stonework invariably calls attention to the German and Czech immigrants who settled there. Russell T. Townsley, publisher of the *Russell Daily News*, mentioned to us that Mrs. Townsley's grandfather, Joseph Riedl, who quarried and cut stone in that area in the last decade of the last century, interestingly enough was born in Austria, his father Ignaz in Berlin, Prussia, and his wife, Mary Deutsch, in Bavaria. On the Ignaz Riedl homestead about eight miles northwest of Hoisington, we examined a small stone house and other stone structures probably built in the homesteading era and a large stone house built in 1911. Joseph (Joe) Riedl's daughter, Mrs. Monte Krug of Great Bend, remembers that although her father assisted with the building of the large home, John Schwartz, a neighbor, quarried the rock (on the quarter section of land on which the house was built) and that Emil Hlavity, who lived a few miles farther west, was the stonemason. Mrs. Joe Riedl, a daughter-in-law of Joe the stonecutter and builder, now lives on the Riedl farm.

Near Olmitz, our attention was directed to a massive barn and a two-story, block-style house of post rock. The house, in excellent condition, is now owned by Mr. and Mrs. Larry Meeks, who are restoring it. Built by a Bohemian stonemason, Erwin Hlavity, in the late 1890s, it has three-foot walls and the limestone blocks have a somewhat unusual design: relatively wide-spaced dents, indicating the spacing of the "teeth" in the tooth-face hammer used by Hlavity to dress the stone. Mrs. Meeks said that she knew of only one other house, also built by Hlavity, having the same design. The quarry from which Hlavity obtained the rock, in a nearby pasture, is now a pond, but the rock at the dam end gives away the secret of its former status.

Travelers who closely observe the post-rock motif along Kansas 4 and 96, as well as along byways in between, from Hoisington to Bazine certainly should enjoy all the more a stop at the Post Rock Museum in La Crosse, at the intersection of Kansas 4 and U.S. 183 near the center of Rush County. In fact Rush County and adjacent northwestern Pawnee and eastern Ness counties might well be an exhibit area for pale, creamy-buff post rock. Immediately northwest of the bridge on Kansas 96 about two miles west of Nekoma, the stone remains of a basement barn in a creek bank direct one's attention to an eleven-room post-rock house on the bank. From correspondence with Clara Williams Humburg of La Crosse, who lived in the house seventeen years, we learned that the original owner of this property, George Abbe, who received his patent in 1885, used one room as a cheese factory. Parenthetically, the house that is now the Post Rock Museum was moved from the Murphy place, located approximately three miles south of this homestead.

Of the numerous stone buildings erected in the vicinity of Bazine and still standing, the Farnsworth barn, on the property owned by Mr. and Mrs. Glenn Schniepp, probably is the oldest. Though the barn has the date 1872 chiseled above its entrance, Mrs. Schniepp thought it probably was not completed until about 1879. The huge structure has several rooms, partitioned by stone walls, and a stone corral is attached. The barn is still in use, though a recent tornado damaged it considerably.

In Ness City on the western periphery of post-rock land, both post rock and Fort Hays chalk have been used for buildings. Several stone buildings remain from the boom period of the 1880s, when as many as forty buildings were under construction at one time, according to Minnie Dubbs Millbrook's history, *Ness, Western County, Kansas.* The best preserved is the Ness County Bank, built in 1887 and now in the National Register of Historic Places. Only part of the building, however, is of post rock. The two walls on the street side, curiously, are predominantly of Cottonwood limestone from Chase County in eastern Kansas. Still standing is the post-rock building that Mrs. S. Mooney mentioned in her letter to the *Plattsburgh* (New York) *Republican,* August 18, 1887: "A fine stone block is just completed called 'Lion Block.' The Lion is carved out of stone [post rock] in a lying posture with head erect, mane flowing and weighs seven hundred pounds."

The first Ness City schoolhouse, built in 1881 and 1882 of Fairport chalk sawed in the form of bricks, is now the Ness County Historical Museum. A well-preserved rural schoolhouse that also was a community building and church and included a cemetery in its yard is the old Buda School in southwestern Ness County. Of the schoolhouse, mostly of Fort Hays chalk but including some Fencepost, Minnie Dubbs Millbrook said, "this Buda community . . . was a God-fearing community and its social and community life centered first at Buda schoolhouse, for a few years at Nonchalanta and then back again to the old Buda schoolhouse. . . . Al-

Front of building of post rock and close-up of lion carved of post rock, Lion Block, Ness City. Architectural design of Lion Block, Romanesque.

though no longer used for school purposes the old stone schoolhouse built in 1884 remains with the cemetery adjoining."

On the Humburg ranch, thirteen miles south and two west of Ness City along Pawnee River, we observed the remains of the old town of Harold, many "modern" stone posts in its fence lines, and an old stone dugout. On the old Thomas Frusher place (now the Robert Frusher place) in northern Hodgeman County, some of the work of that artistic stonemason remains: a stone dwelling and a stone barn; a stone corral with hitching rings along one side; a carved gate archway; flat slabs of rock standing on end to form a shed wall; and a row of tapered posts lining the driveway to the house.

Along Dry Creek about six miles northwest of Hanston, we stopped at the site of the old Mudge Ranch. Margaret Evans Caldwell, Hanston English teacher and author, brought this early ranch to life in her account of the Harvard-educated playboy who lived there between 1880 and 1886. A few stone posts with holes for wooden plugs, a portion of an old stone corral, and (where the house stood) a hand-dug well laid up with native stone are all that remain of a once lavish layout of stonework. Mrs. Caldwell writes that Henry S. Mudge, son of a millionaire woolen manufacturer of Boston, had a huge inheritance. He had studied medicine in this country and abroad and was a musician. Yet he yearned to do some "experimental ranching out West." So he acquired several sections of land in northeastern Hodgeman County and developed what "looked like a young town"—a rambling stone (post-rock) ranch house, a stone stable and corral, several

Buda schoolhouse, predominantly of Fort Hays chalk, in southwestern Ness County.

Tapered posts along the driveway to the Robert Frusher farm (originally owned by Thomas and Helen Frusher), northern Hodgeman County. Thomas Frusher quarried, tapered, and set the posts in a fence around the farm's stone house about 1910; Robert Frusher transferred them to their present position in 1961.

stone and frame sheds and other buildings, not to mention a race track, a polo court, a tennis court, dog kennels, and miles of stone posts. First he raised sheep, then cattle, both haphazardly. Periodically he had wild parties to entertain Eastern friends and sometimes English nobility. In the end his experiment failed. However, Mrs. Caldwell believes the money that he spent in the area, the employment he gave to poverty-stricken homesteaders, and his philanthropic activities boosted the economy of Hodgeman County when a boost was needed most. Frank Salmans of Jetmore now owns the original Mudge acreage, and some of the stone from the old Mudge ranch house has been used in several houses in Jetmore and Hanston.

Though Jetmore is near the southwestern extremity of post-rock country, its downtown area is a good example of a business district built of post

Downtown Jetmore, 1973. Next to the corner building—erected as a bank in 1888—is the T. S. Haun Museum, erected as a combined house and place of business in 1879; limestone blocks from some of the town's dismantled old limestone buildings form the front wall of an addition to the museum.

rock (and Shellrock). Except for the Haun House, erected in 1879, most of the stone buildings in that business block were erected in the late 1880s. About some of those buildings Margaret Haun Raser wrote in 1960, "A man that now owns a building my father built at that time, said they checked it when he bought it, not long ago, and those 70-foot [high] walls were out of line less than a quarter of an inch."

The best time to be in post-rock country, to come under its spell, and to reflect on all that we have seen, from the Republican River to Buckner Creek, is in the early morning or late afternoon. It is then that shading and lighting patterns are most striking. Just after sunrise or just before sunset in the broad expanse of the region, one can look toward the west or toward the east to see rows of stone posts standing and shining, like elongated lights on a landscape awakening for the day or soon to be covered by the night.

7

MAJESTIC AND DURABLE THEY STILL STAND

The 1970s, the decade of the country's bicentennial, coincidentally can be considered as the close of the first century for post rock in use in the Smoky Hills Region of north-central Kansas.

But how much longer will they be here, these graceful stone posts and other stone artifacts that seem to belong in this plains borderland of mid-America? Are they worth preserving—in the only region anywhere that can be called the Land of the Post Rock? Since World War II, or about the time public roads were being widened, the tendency has been to remove the stone posts, and if there are to be replacements, to use wood or steel posts. Many sturdy stone buildings stand abandoned—their fate to become ruins or to be torn down. With that trend, the man-made "finishing touches" on the post-rock landscape may not be a part of the heritage of the region's next generation.

There is an opposite trend, however. Various groups and individuals are involved in efforts to preserve these durable, majestic stone posts and some of their stone counterparts. In the late 1930s travelers, finding their way into the region and asking service station attendants, "What are those stone markers along the roads?" very likely awakened a regional post-rock consciousness.[1]

The man now known to many as "Stone Post" (Ralph) Coffeen was one of the first to be aroused enough to do some "looking into the matter." That was after one day in 1939 when he was accosted by a "lady from

"Over hill and dale" in northeastern Lincoln County.

Boston," who demanded an accurate answer to "What are they?" Being an author, that lady was not satisfied with some of the answers she had been receiving, such as "pebbles grown up" or "Indian grave markers." Until then, Coffeen had taken stone posts for granted. He got some answers for the Bostonian, and since has amassed a lot of information on stone posts, from who quarried them and why to when and how. He has collected post-rock artifacts, he has written letters to his congressmen and senators from Kansas on their preservation, and he has even designed post-rock postcards and stationery.

About that same time stories on or mention of the unusual monuments began appearing occasionally in newspapers and magazines and in books on Kansas. Stone posts are mentioned, for example, in *Kansas, a Guide to the Sunflower State*, published as a Federal Writers' Project in 1939; in *Wheat Country* by William Bracke, 1950; in *This Place Called Kansas* by Charles Howes, 1952; in *The Kansas Scene* by Grace Muilenburg (with photography by Ada Swineford), 1953; and in *Historic Kansas: A Centenary Sketchbook* by Margaret Whittemore, 1954. Cecil Howes, for many years a reporter on the Kansas scene, was one of the first to write an article on stone posts. His story "Early Kansas Settlers Dug Their Fenceposts out of the Ground," much of it based on Kansas Geological Survey reports, appeared in the *Kansas City Star*, April 6, 1940. "The Stone Post of Central Kansas Dates from Pioneers" by Ralph Coffeen appeared in the *Russell Record* May 8, 1941.

Gradually some regional editors began to speak out in defense of stone posts and post rock, some of them vigorously, as did Alan Houghton of the *Beloit Daily Call* in the late 1950s and in the 1960s. Russell T. Townsley of the *Russell Daily News*, who shows his guardian complex by living in a house of post rock, editorialized on the heritage of "the post-rock prairies" when Russell celebrated its centennial. And those coming in from outside needed no coaching to add their support. For example, Robert R. Bruegger, who left the *Hawk Eye* of Burlington, Iowa, to become editor-publisher of the *Hays Daily News* in 1972, said, "The stone fence posts immediately attracted my attention on my first visit to western Kansas and I still feel they're one of the most fascinating sights in the Hays area." In the fall of 1973 Earl and Elaine Loganbill established in Beloit a weekly, *Solomon Valley Post*; a drawing of a corner stone post appears in the masthead.

Increasingly, regional artists, seeing post rock as a characteristic landscape feature, began recording their impressions in sketches or on canvas. Inevitably originals and reproductions began appearing in print or on living room walls or in the halls of public places. For some what began simply as pleasure turned into profit. In 1973 Jane Mehl, of rural Beloit, remarked to us that she has a backlog of requests for paintings of stone posts. Charles B. Rogers has on display in Rogers House Museum-Gallery, Ellsworth, paintings with titles like "Stone Fence Country," which pay homage to post rock. Rare is the resident of artistic bent who does not include stone posts as a subject. For Matilda Altenbaumer Adams, of Lyons, her favorite painting is of a corner stone post, which stirs nostalgia for the Barton County farm where she spent her childhood. When Winstan Anderson, M.D., Lawrence, started painting he revisited his hometown of Tescott, not far from the eastern edge of post-rock country, to seek out as subjects the stone posts he once took for granted.

One of the most visible and positive moves to preserve the post-rock heritage took place in the mid 1950s, when C. C. Abercrombie, Barnard banker, selected Fencepost limestone as the building material for a motel at the intersection of Kansas 18 and 14 north of Lincoln. In 1956 the motel complex (motel, restaurant, service station) was dedicated under the name "Post Rock Motel." The sign bearing that name stands tall on a base of post rock.

Thus, although "post rock" has long been used informally by geologists and others to identify the limestone, it was Abercrombie the banker who coined for the region a "trademark." Since 1956 "post rock" signs have become legion throughout the post-rock area. The term has been incorporated in names for places and projects and in slogans for celebrations and special events; it has appeared in promotional brochures and on labels for souvenirs.[2]

On May 17, 1964, the Post Rock Museum at La Crosse was dedicated, an obvious acknowledgment that the trademark indeed had caught on and should be perpetuated. The museum idea was born in the early 1960s, when some folks in Rush County were having second thoughts about what was happening to their stone posts. As the story goes, thoughts changed to action in 1962 when a customer walked into a La Crosse barber shop and asked, "Why don't you folks 'do something' about those stone posts?" William Appel, the barber, carried the question to a few of his fellow townsmen, and they all said, "Let's do something." The Rush County Historical Society was reactivated to establish and maintain a post-rock museum in the county-seat town of La Crosse. Roy Ehly, manager of the La Crosse Southwestern Bell Telephone Company office, was elected president, and Harry Grass, president of the Farmers and Merchants State Bank, secretary. With donations—$7,515, a site in Grass City Park, and a stone house built in 1883 by Dan Haley near Nekoma—the project "got going." The stone house was moved, stone by stone, to its new location. Then, with Myrl V. Walker as adviser, the Society drew up plans for exhibits, which already were coming in. The exhibits, however, were to be (and are) limited to objects and information that "tell the story of post rock and associated strata of the Greenhorn limestone as they relate to area development." Grass said the intent was (and is) "to preserve what is rapidly disappearing, and to attract tourists."

The featured exhibit in Post Rock Museum simulates an outcrop of Greenhorn limestone showing the Fencepost bed uncovered and partially quarried. Appropriately enough, the Post Rock Museum inspired the establishment of the barbed-wire museum at La Crosse.[3]

Several historical museums in other post-rock towns also contain artifacts and considerable information on post rock. Fossil Station Museum of Russell is one of the largest. Fossil Station, Russell's original name be-

Post Rock Museum, La Crosse.

cause of the "fossils in the rocks" at the site, emphasizes, perhaps, the museum's attention to geology. Jerry Maxfield, a geologist, was its first full-time curator.

The Bunkerhill Historical Museum, opened in 1970, is one of the area's newest museums. It contains among its post-rock artifacts a rock borer, or drilling buck, and the date stone, "J. Winebrunner, 1871, Builder," from the old Harbaugh building, which old accounts refer to as the first business building in Russell County. Torn down only recently, it was in turn an office building, the first Russell County courthouse, a drugstore, a printing shop, a restaurant, a bakery, a cream station, and a residence. The museum building, built of sledged post rock in 1880 as the Lutheran church, is itself a museum piece. A stone sign post stands in the yard, and a flagstone step at the entrance. Members of the Bunkerhill Historical Society living in the area act as curators, donating their services.

In 1964 Jetmore's historic Haun House became the Haun Museum. Erected by T. S. Haun in 1879, it was given to the Hodgeman County Historical Society by his daughter, Margaret Haun Raser. The building is known, also, as the "House of Yesteryear." In a letter written to us in 1960, Mrs. Raser said: "It is still sturdy and strong. No part of it was ever rebuilt. At one time it was rented for a restaurant; the tenant wanted a larger front window. . . . That's the only change." The museum now has an addition, built of native limestone from other old structures that have been torn down in Jetmore and vicinity. Several blocks are date or identification stones. One reads "Public School, 1884." Another reads "Niederacher." Haun House is in the National Register of Historic Places.

In Dodge City, to the south of Jetmore, the Ford County Historical Society maintains the near-century-old Home of Stone, built of post rock from the draws of Sawlog Creek in northern Ford and southern Hodgeman counties. Also known as the Mueller-Schmidt house, it recently became an entry in the National Register of Historic Places because of its "fine quality and contribution to Kansas architecture." Charles L. Hall, in an article in his series on Kansas architecture (*Kansas Country Living*, July 1973), notes that the limestone was "carefully handpicked as to coloring and durability. . . . cut and faced at the quarry before transporting to the site. The mortar was carefully compounded so that the masonry would be durable and long lasting."

The building maintained as a museum by the Jewell Historical Society in downtown Mankato is of post rock, as can be easily determined by examining its front: a wall of two-toned blocks, buff and dark buff (almost brown), characteristic of Fencepost limestone in the northern part of the region.

The Mitchell County Historical Society, though not housed in a building of post rock, nonetheless considers one of its functions to "preserve

The Home of Stone, Dodge City, sponsored by the Ford County Historical Society since 1965. *Photograph by Lola Harper.*

the post-rock heritage," according to Paul V. Grittman of Simpson, Society president. Currently (1973) the Society is remodeling, for its museum, a brick colonial building that for years was the nurses' home at the Mitchell County Hospital. C. R. Hubbard of the Guaranty State Bank said that a sign will be hung between two stone posts, braced as though corner posts. He added that probably a row of stone posts will extend from the gate or entrance along the front of the grounds.

Other county and local historical societies in the region variously keep post-rock awareness alive, if not by being housed in a building of post rock or by having post-rock monuments on museum grounds, then by displaying post-rock artifacts or by maintaining files of old photographs and writings that call attention to post rock's uses in community development.[4]

State and federal agencies responsible for transportation and for recreational facilities in the region show deference to the post-rock tradition. In 1964 the Kansas Highway Commission set stunted stone posts in highway rest areas north of Ellsworth along Interstate 70 (then under construction), and the Kansas State Historical Society erected a historical marker to tell the traveler, "You are now in post-rock country." Later, short posts were set at Interstate rest areas east of Russell and Hays. Still later, stone

Fencepost limestone post supporting water faucet in Minooka Park, south shore of Wilson Reservoir, Russell County.

posts began appearing at service areas along the superhighway in post-rock country.

The Bureau of Reclamation can be credited with salvaging stone posts removed during construction of Glen Elder Reservoir (Waconda Lake), and the U.S. Corps of Engineers for stockpiling posts during construction of Wilson Reservoir. Later the posts were used in landscaping at the recreational sites. Uses at Wilson Reservoir are diverse: stone posts at various places around Wilson Lake and in the park areas; a stone-post support for a drinking faucet, stone-post park benches, and post-rock buildings at Minooka Park; stone corner posts at the dam area, as well as a stand built of post rock to cradle two signs, one for a quote beginning, "This is post-rock country." Salvaging the limestone posts for use at the Wilson area had been recommended strongly by Coffeen. The policy adopted by the U.S. Corps of Engineers is in accord with that recommendation: "Salvage of the posts, which can be accomplished within reasonable cost or economical contract operations, will result in their being stockpiled for later use at recreation or operation areas."

Meanwhile, a few farmers in various localities continue to favor stone posts over new-fangled substitutes. In northern Lincoln County, Golden Morris, old-time quarryman, still maintains his fence lines by stone-post replacements; if any on road embankments begin to lean because of soil creep, he straightens them. Not far from the Morris farm, Earl Keeler of rural Barnard uses stone posts in old and new fence lines. South of Hunter and in the Ash Grove neighborhood, new and old stone posts are a common sight. Arthur Jepsen showed us one of the quarries on his old place east of Ash Grove, where he obtained stone posts for fence lines until recently. On his farm south of Sylvan Grove, Ernest Hanzlicek has removed a few stone posts from where they are no longer needed, but still he prefers stone posts to the modern substitutes. He explained, "Why should I replace a stone post that has served well for seventy years or more with a wooden post that would cost me from $1.75 to $2.75 and probably last only a few years?"

At the Humburg ranch near the Hodgeman-Ness county line, Harold Humburg has set many fence lines in the traditional manner. Harold, who has operated this ranch since 1920, over the years has increased his appreciation of the durability of a stone-post fence. He noted, however, "The advent of the electric fence meant the destruction of the majority of stone post fences in Kansas." To preserve this unique fence and still utilize the advantages of the electric fence, he developed a simple technique to attach electric wire to a stone post.

When we visited the ranch in the 1960s, Gene Humburg showed us the farm quarry that was the source of many of the "new" stone posts we

A house of post rock that reflects Beloit's proud heritage. Erected by C. A. Perdue in 1873 and added to in 1879. Current owners and occupants, Mrs. Margaret Hanni and Miss Jeanne Hanni, have restored the house and placed in its rooms furnishings of appropriate vintage.

saw in the forty-two miles of fence lines then on the Humburg property. In quarrying the posts, Gene and his father, Harold Humburg, use feathers and wedges (as did the pioneers) but electric drills (in place of those powered by hand or by gasoline engines). We were shown several other limestone exhibits on the ranch, which encompasses the townsite of Harold. The Humburgs have restored and modernized the "Dilley" house, built in 1886 by one of the town's founders, M. W. Dilley. Also, we observed a well-preserved stone dugout, the remains of another, and an old well cover fashioned of a slab of native limestone.

That a number of old stone houses throughout the region are being restored and lived in indicates that their charm has carried over to the modern era and that some members of the modern generation appreciate them enough to restore them before they become ruins. That a number of old post-rock buildings in the business districts of many of the region's towns are well preserved and in use indicates community respect for the native limestone. The campus of Fort Hays Kansas State College offers a treasury of preserved structures of native limestone, predominantly post rock. Among the most majestic of the preserved or restored post-rock buildings are the churches, which range from the strikingly simple to those akin to cathedrals.

A stone post, a field, and a "cathedral." Southwest of Pfeifer, Ellis County.

MAJESTIC AND DURABLE THEY STILL STAND

In the outcrop, in fenceposts, in buildings—the two-toned post rock is a top attraction of the Kansas scene and the special heritage of the Smoky Hills Region. As stated in *The Land of the Post Rock* issued in 1956, "To preserve the post-rock landscape is an obligation the people of the area have to themselves, to the state, and to all visitors who come to admire it."

Near Paradise, Russell County.

Notes

CHAPTER 1—A GEOGRAPHIC SETTING

1. The sandstone barn with the post-rock trim in the southeastern corner of Lincoln County is on the Andrew Yordy homestead, owned by Mrs. Lois Muchow of Brookville. Dwight A. Yordy, Andrew's grandson, who lives on a neighboring farm, stated in a letter that James Little built the barn "for the price of a team of horses." Stone for the barn walls, however, was laid by Hans Jorgen (John George Nelson), who had learned the stone masonry trade in his native Denmark, according to a granddaughter, Louise (Mrs. W. Carl) Johnson of Salina. Helen Craig Dingler of Enterprise, who as a child often visited the Yordy farm, wrote that the barn "walls are three feet thick. Thru all these years, the barn has not settled nor sagged. The doors still swing freely."

2. For an overview of regional physiography and topography, we consulted various publications, especially those of the Kansas Geological Survey based on investigations of the post-rock area. Primary sources include "The Kansas Landscape" in *Pleistocene Geology of Kansas* by John C. Frye and A. Byron Leonard; and Walter H. Schoewe's "Physical Geography," part two of a series on Kansas geography published in the *Kansas Academy of Science Transactions.*

3. Those interested in the area's soils should consult the map *Soils of Kansas*, compiled by O. W. Bidwell and C. W. McBee.

4. The generalized discussion of regional geography, climate, and agriculture includes our own observations plus information from recent annual reports of the Kansas State Board of Agriculture. In addition we referred to Snowden D. Flora's *Climate of Kansas* and Glenn T. Trewartha's "Climate and Settlement of the Sub-humid Lands." We consulted various members of the Kansas Agricultural Experiment Station, and for up-to-date climatological data we relied on Merle Brown, who until recently was climatologist for Kansas, National Weather Service.

CHAPTER 2—LAND OF THE POST ROCK ABORNING

1. Origin of "Home on the Range" was not known until after Franklin D. Roosevelt had declared it a favorite of his and it had become one of the country's hit songs in the mid 1930s. More than a few persons claimed authorship, but when the authors of "My Arizona Home," claiming that to be the "parent" song, filed a suit against thirty-five persons and corporations for infringement of copyright, a determined search was made for the song's origin. It was established that Dr. Brewster Higley wrote the words and Dan Kelley, a fellow Smith County pioneer, wrote the music in 1873. Soon afterward the "poem" most certainly was published

as "My Western Home" in the *Smith County Pioneer* (though no copies of the issue in which it appeared are extant). It was republished in the February 19, 1914, issue of *The Pioneer* with this editorial comment, "After a lapse of more than forty years we again offer to the readers of *The Pioneer* the good, old time song as it was first published in 1873." Later the text of the song, with Dr. Higley given as author, was discovered in the February 26, 1876, issue of the *Kirwin Chief.* "Home on the Range" officially became the state song in 1947. (This summary is based on "Home on the Range" by Kirke Mechem.)

2. Information on homesteading is from several sources but principally from a circular issued in 1875 by the United States Department of the Interior and reproduced in the *Fourth Annual Report* of the Kansas State Board of Agriculture for 1875. Homesteading requirements varied among the states, and the original Homestead Act subsequently was amended several times. Roy M. Robbins in *Our Landed Heritage* and Paul W. Gates in *Fifty Million Acres* discuss preemption and home-steading in detail and include extensive references.

3. This summary is an oversimplification of the status of the public domain in Kansas during this period. Paul W. Gates in *Fifty Million Acres* discusses it adequately, with due regard for the Indian problem. (The status of the Plains Indians before and after the government-removal policy is summarized in "Indians in Kansas" by Eugene R. Craine.) The intent here is simply to point up one of the reasons central Kansas was "the place to go" when the Homestead Act became effective.

4. Generally attributed to Greeley, "Go West, young man" was originated by John B. L. Soule in an 1851 edition of the *Terre Haute* (Indiana) *Express,* according to *Bartlett's Familiar Quotations.* Greeley, who credited Soule with the slogan, popu-larized it in his editorials, lectures, and letters. Van Deusen in *Horace Greeley: Nineteenth-Century Crusader* (page 173) notes that "the legend began to grow that Greeley was the originator of the phrase" after Greeley said it to Josiah B. Grinnell in 1853. Stoddard in *Horace Greeley: Printer, Editor, Crusader* points out that Greeley became known as "Go West Greeley" when he was the young editor of the *New Yorker;* every issue of the magazine in 1837 and 1838 urged depression victims to go west.

5. Various accounts of Indian depredations, written from different points of view, appear in regional histories. I. O. Savage in *A History of Republic County, Kansas* generalized about the scares that discouraged settlement in parts of post-rock country: ". . . no part of Kansas suffered more severely from Indian raids and depredations than the Solomon, Republican and White Rock Valleys." Harry E. Ross in *What Price White Rock?* (pages 5 through 26) and M. Winsor and James A. Scarbrough in "Jewell County" provided details, not without empathy for the settlers, though Ross included some explanation for the motives of the Indians. Shying from embellishments, C. Bernhardt in *Indian Raids in Lincoln County, 1864 and 1869* stated in his preface, "public records have been my guide, supplemented by such information as can be had from the pioneers and scouts."

6. In characterizing *The Great Plains,* Walter Prescott Webb considers and places in perspective most aspects that influenced the area's settlement and development: the natural environment, including especially the scarcity of timber and water; the culture of the Plains Indians; and the evolution of the range-cattle industry, treated in association with the coming of the railroads and the fencing problem.

7. Cattle from the Texas ranges posed a greater threat to settlers in the southern part of post-rock country than to those in the northern part, most of which was quaran-tined against entrance of Texas cattle. The influx of Texas cattle, however, provided

some advantages to some, at first. A settler might acquire his first stock as culls from the range herds, sell provisions to the herdsmen, or even "winter" some of the cattle. Francis J. Swehla in "Bohemians in Central Kansas" alludes to both anti and pro sentiment toward Texas drovers.

Those curious about the nature of the Texas cattle drives, including life along the trails and in the cow towns, and about the range-land controversy, including legal battles between cattlemen and homesteaders, will find these sources rewarding: Edward Dale's *Range Cattle Industry*, Jimmy M. Skaggs's *Cattle-Trailing Industry: Between Supply and Demand, 1866–1890*, and Robert W. Richmond's "Cowtowns and Cattle Trails" (emphasis on Kansas). For added incidents of human interest, including reproductions of trail logs and cowboy songs, read "The Long Drive" by Everett Dick.

8. Two works of J. Neale Carman, *Foreign-Language Units of Kansas* and "The Foreign Mark on Kansas," can be considered nuclei for discussions on European immigrants in post-rock country. Major general sources include chapter 14 ("Who Are These Jayhawkers?") of Zornow's *Kansas* and chapter VII ("Luring the European") of *Our Landed Heritage* by Robbins. Sources for specific settlements were legion.

9. Details on the Excelsior Colony were hard to come by. We found no one source that gave a lucid, complete accounting of all divisions. We pieced together partial, and to some extent conflicting, accounts from several sources, chiefly "Jewell County" by Winsor and Scarbrough, *What Price White Rock?* by Ross, and *The History of Republic County* by Blackburn and Cardwell. Interviews and correspondence with Mr. and Mrs. George Doctor of Belleville enabled us to draw conclusions with confidence.

10. Many articles, containing various assertions and interpretations, have been written about the British experiment at Victoria. For our summary account we relied primarily on two sources that we consider credible: Marjorie Gamet Raish's *Victoria: The Story of a Western Kansas Town*, based on grass-roots research; and correspondence with and a brochure (*George Grant, the English Colony, the Villa*) by William Baier, who owns the Grant estate south of Victoria and who for many years lived in the villa.

11. Father Laing's work, our basic published reference on the German-Russians—or Russian-Germans or Volga Germans—in the Hays vicinity was compiled largely from manuscripts, diaries, and family books written in German by original settlers. The most comprehensive was a hand-written, 600-page manuscript by Joseph Linenberger, Sr., the leader of a major group that came to Herzog (Victoria) in September 1878. Herman Linenberger, whom we interviewed in Victoria, is a son.

Conquering the Wind by Amy Brungardt Toepfer and Agnes Dreiling should be mentioned as another important source for tracing the background of these peoples, beginning with their life in Germany, continuing through their century on the meadows and steppes of the Volga region of Russia, and culminating in their coming to the plains of Ellis and Rush counties, Kansas.

CHAPTER 3—SAGA OF THE STONE POSTS

1. North-central Kansas counties affected by the 1870 herd law included Saline, Ottawa, Washington, and Cloud; Republic County was included in the 1871 law. The law of 1872, printed in the official state newspaper, the *Topeka Commonwealth*, on February 25, 1872, could be called into force the following April 6. The *Fourth*

Annual Report of the State Board of Agriculture, on pages 498 to 502, interprets the herd law and also the fence laws of Kansas in force at that date. Cost of fencing in the United States is discussed, and the common law on fencing—that the owner of real estate is entitled to the exclusive possession of this property—is explained in the context of the fence laws of Kansas: "The effect of legislation of Kansas, so far as it modifies the common law in relation to fences is as follows: Unless a party shall maintain a lawful fence, he does not take such care of his own land and crops as to enable him to recover damages which might have been avoided had he kept up a good fence."

Rodney O. Davis, in an unpublished thesis, "The Fencing Problem and Herd Law in Kansas 1855–1883," discusses in some detail the factors involved in the agitation for herd laws: the Texas cattle trade, conflicts between stockmen and cultivators of the soil, lack of fencing materials, and initially the high cost of installing a legal wire fence around 160 acres (at least $384—more than most homesteaders could afford); the effectiveness of the herd law of 1872, including public reaction to it; and finally the lack of need for it as cheap and readily available barbed wire came into use (here assuming post timber or a suitable post substitute also was available).

2. Also called Texas or Spanish fever, the disease attacked the spleen and eliminative systems, resulting in blindness and death for the northern cattle; it seldom affected the Texas cattle in that they had developed an immunity. Having observed that after they had wintered in the state Texas cattle no longer carried the fever, some Kansas settlers were willing to sell provisions and winter feed and to rent pasture to the herdsmen; some even purchased Texas stock. But acceptance was only temporary. As stated by Davis ("The Fencing Problem and the Herd Law in Kansas, 1855–1883"), "Resentment remained strong against fever-bearing Texas animals wherever they posed a threat to native stock, and bad feeling toward all Texas cattle was compounded as greater numbers of purebred shorthorn and other high-grade stock were imported, and as cereal culture developed on the central Kansas prairie."

Late in the territorial period and during early statehood, the Kansas legislature enacted a series of laws aimed at restricting Texas cattle by area and season. An act of 1867 "unconditionally" prohibited Texas cattle between March 1 and December 1 of each year "from that portion of the state east of the sixth principal meridian; also to all that portion of the state north of township nineteen (19), but shall not apply to that portion of southwest Kansas west of the sixth meridian and south of township eighteen (18) . . . [but] all persons herding or driving the description of the stock mentioned in the first section of this act in that portion of Kansas west of the sixth principal meridian and south of township eighteen (18), within five miles of any public highway, or any ranche or other settlement, without the consent of the settler or owner of such ranche, shall be liable . . ." (*General Statutes of Kansas*, 1868, chapter 105, section 12, page 1016). As farmsteads, and protests, increased in central Kansas, that "Texas quarantine line" (meaning the line separating the area from which Texas cattle were prohibited at all times from that where they were not) was moved westward; consequently, cow towns also shifted westward. In "Cowtowns and Cattle Trails," Robert W. Richmond brings that aspect of the fencing drama into focus.

3. The Kansas legislature of 1883 not only made barbed wire legal fencing material but also repealed the hedge bounty law. Even before its legal sanction, barbed wire was increasing in use. Whereas the *First Biennial Report* (1877–78) of the State

Board of Agriculture stated that Kansas had only 1,700,000 rods of wire fence, the *Third Biennial Report* (1881–82) reported 10 million rods. Though smooth and barbed wire were not distinguished, most probably was barbed. Also, by the 1880s advertisements had begun to appear in newspapers read by prairie farmers in a mood to try any fencing material that was available and that they could afford.

4. Minnie Dubbs Millbrook, Kansas historian and author of *Ness: Western County, Kansas*, made available to us transcripts of letters that Mrs. Seymour Mooney, early Ness County settler, wrote to the *Plattsburgh* (New York) *Republican*.

5. Hubert Risser in *Kansas Building Stone* (pages 64 and 65) states: "Most stone when freshly quarried contains a certain amount of moisture, or sap, which is gradually lost as the stone is exposed to the air. . . . Many stones tend to 'season' or 'case-harden' during the drying process that takes place after quarrying. . . . One of the most commonly accepted explanations is that quantities of the same kind of materials that form the binder cementing the individual grains of the stone together are held in solution in the water occurring naturally within the stone. When the freshly quarried stone is exposed to the air, the solution is drawn to the surface through capillary action, and the water evaporates, leaving the cementing material behind to form a protective coating at the surface of the stone. The weather-resistance of stone seems to be increased by this process."

CHAPTER 4—POST ROCK'S USES IN PERSPECTIVE

1. During the last twenty to thirty years of the nineteenth century, limestones of the Greenhorn, Niobrara, Ogallala, and other formations (including Permian limestones) in Kansas were referred to locally as "magnesian limestone"; sometimes even today in central Kansas the term is heard. It appears in many of the county summaries in the old annual and biennial reports of the Kansas State Board of Agriculture (1872 to the mid 1890s), but the specific meaning is not clear. Kansas Cretaceous limestone contains only small quantities of magnesium compared with other limestone. There are hints that "magnesian" was used to refer to relatively soft limestone, though we find references to "hard magnesian limestone."

 One description suggests that color was a factor. The *Third Biennial Report* has this to say: "Blue, white and magnesian limestone is found in several localities [in Rawlins county]."

 In the selfsame volumes, sections written by geologists either avoided the term entirely or criticized its use. Thus, as early as 1878, B. F. Mudge in "Geology of Kansas" *(First Biennial Report)* writes: "Some of the [Permian] lime has been called magnesian, but analysis has failed to show, in more than a single instance, over five per cent. of magnesia." Geologist Robert Hay in "Northwest Kansas: Its Topography, Geology, Climate, and Resources" *(Sixth Biennial Report)* hints that perhaps the term never did refer to the element magnesium. He writes of "The Niobrara limestones which are locally called magnesian (or rather *magneezia*)." In the same report, the Riley County summary refers to the Permian-age Fort Riley limestone as "a buffy, soft, magnesian limestone, with very little magnesia."

 In 1898, Erasmus Haworth *(Annual Bulletin on Mineral Resources of Kansas for 1897)* summarized the situation: "In the central and west-central part of the state, the Cretaceous limestones have been quarried to a great extent. On account of their soft, chalky character, they are generally spoken of locally as a magnesian limestone, although such a term is entirely misapplied."

2. A cord of stone would equal 128 cubic feet in a stack measuring 4 by 4 by 8 feet. Var-

ious measurements, however, were used. Commonly, stone was sold by the perch: 24.75 cubic feet, usually in units measuring 16.5 feet (1 rod) by 1 foot by 1.5 feet. Stone also was sold by the ton, by the square foot, by the cubic foot, and even by the block (the price depending on the size of the block) or by the wagon load.

3. The Post Rock Museum, La Crosse, has on display many tools and devices used at various times. Among them are several types of stone hammers; pitch-face, smooth-face, and hammer-face chisels; a stone crandall (a large hammer used in dressing and shaping stone); stone drills and drilling bucks; jointers (bent pieces of iron inserted to strengthen joints); tying pins (used in mortar joints); and of course feathers and wedges. One exhibit is a piece of iron, which (when bolted to an anvil) was used to shape molten pieces of metal into feathers by pounding them into an indentation on one end of the piece of iron. Another exhibit is a 1930-style drilling machine, complete with its Maytag gasoline motor, made by Andrew Herrman of Liebenthal.

4. Definitions of some of the terms used in the building-stone trade may be useful to readers. The terminology given here is from Risser's *Kansas Building Stone* bulletin, except as noted.

Ashlar. Small, rectangular blocks of stone having sawed, planed, or rough-faced surfaces. Blocks may be of uniform length but commonly are of random length. They may be laid in even-height course pattern, in which all blocks of any course are of equal height, or in a random pattern in which blocks of two or more heights are fitted into a pattern. In *random pattern* the small blocks are laid so that their combined height equals that of the larger blocks in a course. Some ashlar stone is cut on all six sides; generally, however, the exposed side has a *broken (split) face*, which is rough though fairly straight, or the edges have been chiseled back to leave a *pitched* (protruding) face; a *chipped face* has an "in-between" roughness. Back and ends of the blocks may be rough or sawed.

Cut stone. Blocks that are carefully cut to a particular shape and size so that they will fit into specified places in the structure. (Ashlar blocks, which may be cut on all six sides, are not finished or accurately sized as are blocks of cut stone.)

Flagging, flagstone. Flat slabs, or flags, sawed from blocks of stone or obtained by splitting slabs from thin-bedded rock layers. Though the terms *flagging* and *flagstone* often are used synonymously, the American Geological Institute *(Dictionary of Geological Terms)* makes this distinction: *Flagstone* is a rock that splits readily into slabs suitable for flagging. *Flagging* is used principally for sidewalks and floors but may be used as veneer on exterior walls.

Quarry face. The freshly split face of ashlar, squared off for the joints only, as it comes from the quarry, and used especially for massive work *(Dictionary of Geological Terms,* American Geological Institute). Blocks used as they come from the quarry commonly are called simply *quarry stone.*

Rough building stone. Blocks of various shapes and sizes. The blocks are hewn in rough shapes and fitted into the wall to form irregular joints, without definite design or pattern. Construction using *rough-hewn* stone was common in former years.

Rubble. Fragments of irregular shape and size. Many such fragments have one good face and are laid in the wall so that the good face shows. L. W. Currier *(Geologic Appraisal of Dimension-Stone Deposits)* points out that the terms "ashlar" and "rubble" are sometimes confused because some producers apply the term "ashlar" to completely squared blocks and others include blocks with two parallel faces but unsquared ends. All ashlar, however, lends itself to laying in courses, whereas rubble does not.

Shiners. Blocks of stone laid in a wall perpendicular to the natural bed. It has been found that stone laid in that manner offers much less resistance to weathering than that laid on the bedding plane.

Veneer slabs that simulate the appearance of ashlar and rubble walls commonly are used in modern construction. L. W. Currier in *Geologic Appraisal of Dimension-Stone Deposits* states: "In the ashlar-type veneer both coursed and random rectangular faces are used; in the rubble type the exposed faces are polygonal but are not sized or accurately shaped, giving what is known as spider web veneer. Both ashlar and rubble veneer blocks are . . . prepared to specified thicknesses, usually less than 3 inches."

5. In a letter dated February 5, 1973, Ballou stated that he molded the bas-relief from a clay image and poured concrete over it, adding that "cement coloring affects the gray of concrete rendering the relief in a color similar to the post rock." By profession a newspaperman (he first was with the *Delphos Republican* and then with the *Salina Union*, which merged with the *Journal* in 1925), Ballou has two major hobbies: painting—one of his favorite subjects is stone posts—and laying stone.

6. Kirk Raynesford wrote in a letter dated January 27, 1973, that he helped his father with the surveys of the Butterfield stations and drew the maps (now in the museum maintained by the Butterfield Trail Historical Association, Russell Springs, Kansas). He noted that his father walked and mapped the BOD Trail through Ellsworth, Russell, Ellis, Trego, Gove, and Wallace counties. (That part of the Atchison-to-Denver route was along the Smoky Hill Trail proper; the eastern part commonly was referred to as the "military road.") We read from material prepared by Howard C. Raynesford on the "Butterfield Overland Despatch" for Russell's *Prairiesta-100*: "David A. Butterfield took charge of the route in 1865. . . . The first wagon train —'Train A'—on June 4, 1865, loaded with 150,000 pounds of freight for Denver, was followed by the first coach which arrived in Denver on September 23, 1865. . . . The rate of travel by the ox team was about twelve to fourteen miles a day, making a trip of 45 to 55 days continuous travel. . . . The Indians resented this invasion of their favorite hunting ground . . . destroyed many of the stations and wagon trains and killed many pioneers, despite the several locations of U.S. cavalry and companies of infantry stationed along the trail."

CHAPTER 5—WHENCE THE FENCE: GEOLOGY OF POST-ROCK COUNTRY

1. The sources of information for our discussion of the geology of post-rock country are so numerous that we can name here only a few. Besides conversations with colleagues Myrl V. Walker and Donald E. Hattin, we consulted publications of early workers but relied heavily on those of several modern investigators, especially Hattin, J. B. Reeside, Jr., W. W. Rubey, N. W. Bass, John Frye, and A. B. Leonard. Some residents of the area, particularly Harvey Roush of Lincoln, guided us to unusual outcrops.

   Some readers may wish to study the geology of the region in greater detail. For a study of the Cretaceous rocks of central Kansas the interested reader is referred to Hattin's 1965 GSA Field Conference Guidebook. The same book includes a graphic column of Greenhorn strata that, along with the tabular descriptions of the various members of the Greenhorn and adjacent formations, should make it possible to identify specific outcrops.

CHAPTER 6—ON A JOURNEY THROUGH THE LAND OF THE POST ROCK

1. Though Greenhorn limestone, including the Fencepost bed, extends into Nebraska, in that state it is largely covered by silt and other Quaternary deposits, except near the Kansas border generally east of U.S. 81 and west of Kansas 15. That cover, along with the topography and the availability of timber and other material, probably deterred the quarrying of the Fencepost bed for fence posts and building stone, according to recent correspondence with Marvin P. Carlson of the University of Nebraska—Lincoln, Conservation and Survey Division.

2. Walter H. Schoewe on pages 102 and 103 of *Coal Resources of the Cretaceous System* gives the mining history of the Minersville lignite district, including a map locating the mines (some of which produced coal as late as 1940). He points out that mine shafts were about twenty-five feet deep in 1875 but that the last mines opened were about one hundred feet deep. The coal was hauled away by the wagon load to as far as twenty or thirty miles, and miners were paid from $1.25 to $2.50 a ton for mining the coal (lignite), which sold for $2.25 to $4.00 a ton. Schoewe notes, "Minersville, now entirely abandoned [1952], was a mining community of several houses, a hotel, stores, and a U.S. post office." Agnes Tolbert describes some of the buildings in *The Rock Houses of Minersville*.

3. The story of the town's first hotel, a two-story stone building known as the "Wilson House," is recounted by Rosanna Healey in "Sketching Early Wilson" in the *Wilson World*, July 15, 1948. It was completed in 1875 by John Jellison, who then traded it for a farm. Isaac Wilson came from the East on a preacher's railroad pass given to him by Jellison to take over and open the hotel for business. Later, presumably "twelve or thirteen covered wagons of Wilsons" came to Bosland (Wilson) and lived in the hotel. Railroad brakemen and engineers, who often ate and slept there, "came to know it as the Wilson House." Ms. Healey, in the article (one in a series of the same title), suggests that an area horse trader by the name of Wilson, rather than the occupants of "Wilson House," probably gave the town its name (when changed from Bosland).

4. The term "Benton" is a relic of nineteenth-century pioneer geologic investigations in the Great Plains. Cretaceous rocks of Kansas were subdivided into three groups: Dacotah (Dakota), Benton (Fort Benton), and Niobrara. (See, for example: Mudge, "Geology of Kansas"; and Hay, "Northwest Kansas" and "Geology and Mineral Resources of Kansas.") The Benton later became the Graneros, Greenhorn, and Carlile formations; the overlying Niobrara was subdivided into Niobrara chalk and Pierre shale. Although geologists stopped referring to Benton rocks many decades ago, the term "Benton limestone" is still sometimes used in post-rock country to refer to the Fencepost bed and associated limestones. That is especially so in the Hays vicinity. Building stone quarried from these beds has been marketed under that name, for example, by George J. Klaus, local quarrier and stone cutter, who for many years has processed local stone (and imported as well) in his shop, Hays Cut Stone and Veneer, on old U.S. 40 west of town.

CHAPTER 7—MAJESTIC AND DURABLE
THEY STILL STAND

1. As quotes and citations in previous chapters might indicate, Kansas Geological Survey reports based on geologic investigations in post-rock counties are to date the best sources of published information on the occurrence and nature of post rock in relation to its uses. Especially useful are accounts in the annual reports on Kansas mineral resources prepared by Erasmus Haworth in the late 1890s and early 1900s and in reports issued between 1924 and 1933 for these counties: Russell, 1925 (Rubey and Bass); Ellis, 1926 (Bass); Mitchell and Osborne, 1930 (Landes); Cloud and Republic, 1930 (Wing); and Ness and Hodgeman, 1932 (Moss). Characteristics and uses of Fencepost limestone also are mentioned in area county reports published later: Rush, 1938 (Landes and Keroher); Ford, 1942 (Waite); Republic and northern Cloud, 1948 (Fishel); Pawnee and Edwards, 1949 (McLaughlin); Barton and Stafford, 1950 (Latta); Lincoln, 1952 (Berry); Jewell, 1955 (Fishel and Leonard); Cloud, 1959 (Bayne and Walters); Mitchell, 1959 (Hodson); southern Ellis and parts of Trego and Rush, 1961 (Leonard and Berry); Ottawa, 1962 (Mack); Ellsworth, 1971 (Bayne, Franks, and Ives); and Rush, 1973 (McNellis). Hattin's work on the Cretaceous for the Geological Survey is an especially valuable source.

   Considerable information from the above-mentioned Geological Survey reports published before 1956 appears in quoted form or is noted in the mimeographed version of *Land of the Post Rock* (1956, 1958), which in turn became a source of information for newspaper stories in regional newspapers and occasionally elsewhere (for example, "Kansas Fenced in by Personalities of Stone" by Susan Marsh, *New York Times*, July 12, 1964).

2. Unfortunately, some "post-rock" souvenirs are not of post rock but of synthetic material. Artifacts of post rock (or associated limestones) that we have seen include miniature stone-post paperweights made by George Welling of Paradise and miniature corner posts produced by Wayne Naegele of Lucas. About the time the Post Rock Museum at La Crosse opened, Harry Grass, secretary of the Rush County Historical Society, had a hobby of making scale-model corner posts.

3. Soon after Leo Schugart of Hoisington, "dean of barbed wire collecting," placed a panel of barbed wire on temporary display in the Post Rock Museum in 1965, La Crosse became the "Barbed Wire Capital of the World." Roy Ehly, president of the Rush County Historical Society and vice president of the Kansas Barbed Wire Association, said that when Don Wigington of Quinter and about a dozen other Kansas barbed-wire enthusiasts noticed the display, they met in La Crosse where they organized and thus started wire collecting as a hobby. Ivan Krug, La Crosse lawyer, currently is president of the Association and also was the first president. Two days each year—the first Saturday and Sunday in May—collectors from thirty states meet in La Crosse for a barbed-wire-swap-and-sell session.

   The Association-backed Barbed Wire Museum in downtown La Crosse is a repository for more than four hundred types of barbed wire. Mrs. Harry Grass, who explained the Museum exhibits to us, noted that there are probably more than a thousand different types of barbed wire, including at least 622 registered patents. She said that to be of value a wire strand must be at least eighteen inches long and that prices for single strands range from fifty cents to $500. The Museum has exhibits of wire stretchers, wire cutters, wire splicers, post-hole diggers, and other items needed in the "fencing business."

4. The Republic County Historical Society contributed to post-rock preservation by including in a centennial history *(History of Republic County, 1864 to 1964)* many references to the county's stone buildings and to the masonry skills of the British, Scandinavian, and Czech immigrants who settled in that county. The Ottawa County Historical Society endorsed the idea of preservation by sponsoring the construction of the Pike Monument (of post rock and Shellrock) near Delphos. The Lincoln County Historical Society paid homage to its native limestones by reissuing Elizabeth Barr's *Souvenir History of Lincoln County* (first published in 1908), which through historical narrative points up the importance of the limestones in the county's development.

Bibliography and Sources

BOOKS AND ARTICLES

Adams, F. G. 1873. *The Homestead Guide, 1873: Describing the Great Homestead Region in Kansas and Nebraska*. Waterville, Kansas: F. G. Adams.

Andreas, A. T. 1883. *History of the State of Kansas*. Chicago: A. T. Andreas.

Baier, William. n.d. *George Grant, the English Colony, The Villa*. Pamphlet.

Barnhardt, C. 1910. *Indian Raids in Lincoln County, Kansas, 1864 and 1869*. Lincoln, Kansas: The Lincoln Sentinel Print.

Barr, Elizabeth N. 1908. *A Souvenir History of Lincoln County, Kansas*. Reprint. Lincoln, Kansas: Lincoln County Centennial Committee, 1961.

Bartlett, John. 1968. *Familiar Quotations: A Collection of Passages, Phrases, and Proverbs Traced to Their Sources in Ancient and Modern Literature*. 14th ed. revised and enlarged. Edited by Emily Morison Beck. Boston: Little, Brown & Co.

Bass, N. W. 1926. *Geological Investigations in Western Kansas: 1. Geology of Ellis County*. Kansas Geological Survey, Bull. 11:11–52.

Baughman, Robert W. 1961. *Kansas in Maps*. Topeka: Kansas State Historical Society.

Bayne, Charles K.; Franks, Paul C.; and Ives, William Jr. 1971. *Geology and Ground-water Resources of Ellsworth County, Central Kansas*. Kansas Geological Survey, Bull. 201.

Bayne, Charles K., and Walters, K. L. 1959. *Geology and Ground-water Resources of Cloud County, Kansas*. Kansas Geological Survey, Bull. 139.

Bergin, Alfred. 1910. "The Swedish Settlements in Central Kansas." *Collections of the Kansas State Historical Society (1909–1910)* 11:19–46.

Bergman, Denzil Wallace. 1950. "The Greenhorn Limestone in Kansas." Master's thesis, Kansas State University, Manhattan.

Bernard, William R. 1906. "Westport and the Santa Fe Trade." *Collections of the Kansas State Historical Society (1905–1906)* 9:552–578.

Berry, Delmar. 1952. *Geology and Ground-water Resources of Lincoln County, Kansas*. Kansas Geological Survey, Bull. 95.

Bidwell, O. W., and McBee, C. W. 1973. *Soils of Kansas*. Manhattan: Kansas Agricultural Experimental Station. Map.

Blackburn, Anona Shaw, and Cardwell, Myrtle Strom. 1964. *History of Republic County, 1868–1964*. Belleville, Kansas: The Belleville Telescope.

Blackmar, Frank W. 1906. "The History of the Desert." *Transactions of the Kansas State Historical Society (1905–1906)* 9:101–114.

Blackmar, Frank W. 1912. *Kansas: A Cyclopedia of State History, Embracing Events, Institutions, Industries, Cities, Towns, Prominent Persons, Etc.*, 2 vols. Chicago: Standard Publishing Co.

Bracke, William B. 1950. *Wheat Country*. Edited by Erskine Caldwell. New York: Duell, Sloan & Pearce.

Burchett, R. R.; Dreezen, V. H.; Reed, E. C.; and Prichard, G. E. 1972. *Bedrock Geologic Map Showing Thickness of Overlying Quaternary Deposits, Lincoln Quadrangle and Part of Nebraska City Quadrangle, Nebraska and Kansas*. U.S.

Geological Survey, Misc. Geological Investigations Map 1–729. Washington, D.C.: U.S. Government Printing Office.

Byers, O. P. 1928. "When Railroading Outdid the Wild West Stories." *Collections of the Kansas State Historical Society (1926–1928)* 17:339–348. Reprinted from *The Union Pacific Magazine* October 1926.

Caldwell, Margaret Evans. 1958. "The Mudge Ranch." *Kansas Historical Quarterly* 24:285–304.

Carman, J. Neale. 1961. "The Foreign Mark on Kansas." *Journal of the Central Mississippi Valley American Studies Association* 2:66–79.

Carman, J. Neale. 1962. *Foreign-Language Units of Kansas: I. Historical Atlas and Statistics.* Lawrence: The University of Kansas Press.

Choitz, John F. 1967. *A History of Ellsworth County: 1854–1885.* Ellsworth, Kansas: Ellsworth County Historical Society.

Christensen, Thomas Peter. 1928. "The Danish Settlements in Kansas." *Collections of the Kansas State Historical Society (1926–1928)* 17:300–305.

Coffeen, Ralph. 1941. "The Stone Post of Central Kansas Dates from Pioneers." *Russell Record* May 8.

Cooper, F. A. 1958. Cartoon 1107. "It Happened in Kansas." Published in Kansas newspapers.

Cragin, F. W. 1896. "On the Stratigraphy of the Platte Series or Upper Cretaceous of the Plains." *Colorado College Studies* 6:49–52.

Craine, Eugene R. 1956. "The Indians in Kansas." In *Kansas: The First Century,* ed. John D. Bright, 65–89. New York: Lewis Historical Publishing Co.

Cruise, John D. 1910. "Early Days of the Union Pacific." *Collections of the Kansas State Historical Society (1909–1910)* 11:529–549.

Currier, L. W. 1960. *Geologic Appraisal of Dimension-stone Deposits.* U.S. Geological Survey, Bull. 1109. Washington: United States Government Printing Office.

Dale, Edward Everett. 1960. *The Range Cattle Industry: Ranching on the Great Plains from 1865 to 1925.* Norman: University of Oklahoma Press.

Davis, Rodney O. 1959. "The Fencing Problem and the Herd Law in Kansas." Master's thesis, University of Kansas, Lawrence.

Dick, Everett. 1928. "The Long Drive." *Collections of the Kansas State Historical Society (1926–1928)* 17:27–97.

Dictionary of Geological Terms. 1960. 2nd ed. Prepared under the direction of the American Geological Institute, with A. C. Trowbridge as editor. Garden City, New York: Doubleday & Company, Inc.

[Dreiling, B. M.]. [1926?]. *The Golden Jubilee of the German-Russian Settlements of Ellis and Rush Counties, Kansas, 1926.* Hays: Ellis County News.

Federal Writer's Project of the Works Projects Administration. 1939. *Kansas, A Guide to the Sunflower State.* New York: The Viking Press.

Fishel, V. C. 1948. *Ground-water Resources of Republic County and Northern Cloud County, Kansas.* Kansas Geological Survey, Bull. 73.

Fishel, V. C., and Leonard, A. R. 1955. *Geology and Ground-water Resources of Jewell County, Kansas.* Kansas Geological Survey, Bull. 115.

Flora, Snowden D. 1948. "Climate of Kansas." *Report of the Kansas State Board of Agriculture* 57:1–320.

Frye, John C., and Leonard, A. Byron. 1952. *Pleistocene Geology of Kansas.* Kansas Geological Survey, Bull. 99.

Garred, Willard R., ed. 1953. *The Evangelical Lutheran Community Church of Denmark, Kansas: Diamond Jubilee.* Salina, Kansas: Arrow Printing Co.

Gates, Paul Wallace. 1954. *Fifty Million Acres: Conflicts over Kansas Land Policy, 1854–1890.* Ithaca: Cornell University Press.

Geologic Map of Kansas. 1964. Prepared by Geological Survey staff under direction of J. M. Jewett. Kansas Geological Survey. Map M-1.

Gilbert, G. K. 1896. *The Underground Water of the Arkansas Valley in Eastern Colorado.* U.S. Geological Survey, 17th Annual Report, part 2:551–601.

Goodheart, Mrs. Ralph. 1962. "A Partial History of the Elm Creek Community." *Russell County Record* January 15.

Hall, Charles L. "Historic Architecture in Kansas." Series appearing in *Country*

Living: June 1971 (Cathedral of the Plains); January 1972 (Skyscraper of the Plains); October 1972 (The Hart residence), December 1972 (The T. S. Haun house); April 1973 (George Grant and his villa); July 1973 (The Home of Stone).

Haney, E. D. 1928. "The Experiences of a Homesteader in Kansas." *Collections of the Kansas State Historical Society (1926–1928)* 17:305–325.

Hattin, Donald E. 1962. *Stratigraphy of the Carlile Shale (Upper Cretaceous) in Kansas.* Kansas Geological Survey, Bull. 156.

Hattin, Donald E. 1964. "Cyclic Sedimentation in the Colorado Group of West-central Kansas." In *Symposium on Cyclic Sedimentation,* ed. D. F. Merriam. Kansas Geological Survey, Bull. 169:205–217.

Hattin, Donald E. 1965. *Stratigraphy of the Graneros Shale (Upper Cretaceous) in Central Kansas.* Kansas Geological Survey, Bull. 178.

Hattin, Donald E. 1965. *Upper Cretaceous Stratigraphy, Paleontology, and Paleoecology of Western Kansas.* Geological Society of America, Field Conference Guidebook.

Haworth, Erasmus. 1898. *Annual Bulletin on Mineral Resources of Kansas for 1897.* University Geological Survey of Kansas.

Haworth, Erasmus. 1902. *Annual Bulletin on the Mineral Resources of Kansas for 1900 and 1901.* University Geological Survey of Kansas.

Hay, Robert. 1889. "Northwest Kansas: Its Topography, Geology, Climate, and Resources." *Kansas State Board of Agriculture Sixth Biennial Report* 92–116.

Hay, Robert. 1893. "Geology and Mineral Resources of Kansas." *Kansas State Board of Agriculture Eighth Biennial Report* 91–166.

Hayter, Earl W. 1939. "Barbed Wire Fencing—A Prairie Invention." *Agricultural History* October 1939. Reprinted in *Heritage of Kansas* 4 (no. 3): 7–16, 1960.

Healey, Rosanna. 1948. "Sketching Early Wilson." *Wilson World* July 15 and 29.

Higley, Brewster. 1873. "Home on the Range." First published as "My Western Home" in the *Smith County Pioneer* (1873); reprinted in that paper February 19, 1914. Also printed in the *Kirwin Chief* February 16, 1876. (See Mechem, Kirke.)

Historical and Descriptive Review of Kansas: The Northern Section. 1890. Topeka: Jno. Lethem.

Hodson, Warren G. 1959. *Geology and Ground-water Resources of Mitchell County, Kansas.* Kansas Geological Survey, Bull. 140.

Howes, Cecil. 1940. "Early Kansas Settlers Dug their Fenceposts out of the Ground." *Kansas City Star* April 6.

Howes, Charles C. 1952. *This Place Called Kansas.* Norman: University of Oklahoma Press.

Hughes, Harry, and Dingler, Helen Craig. 1973. *Picture Trails Past to Present.* Enterprise, Kansas: H & D Press.

Hull, O. C. 1912. "Railroads in Kansas." *Collections of the Kansas State Historical Society (1911–1912)* 12:37–52.

J[acquart?], R[olland?] R. 1928. "Hard-riding Red Coats Put Gay Touches on Kansas Plains." *Kansas City Star* March 11.

Jelinek, George. n.d. "Fencing: Facts and Fancies." Mimeographed.

Jensen, Burnice Pearson. 1971. "History of Saron Baptist Church." In centennial pamphlet (1871–1971). Mimeographed.

Jewett, J. M. 1961. "The Geological Making of Kansas." *Kansas Academy of Science, Transactions* 64:175–197.

Jewett, J. M.; Bayne, C. K.; Goebel, E. D.; O'Connor, H. G.; Swineford, Ada; Zeller, D. E. 1968. *The Stratigraphic Succession in Kansas.* Kansas Geological Survey, Bull. 189.

Kansas State Board of Agriculture. 1874. *Report for the Year 1873.*

Kansas State Board of Agriculture. 1874. *The Third Annual Report, for the Year 1874.*

Kansas State Board of Agriculture. 1875. *Fourth Annual Report, for the Year Ending November 30, 1875.*

Kansas State Board of Agriculture. 1878. *First Biennial Report, for the Years 1877–8.*

Kansas State Board of Agriculture. 1881. *Second Biennial Report, for the Years 1879–80.*

Kansas State Board of Agriculture. 1883. *Third Biennial Report, for the Years 1881–82.*

Kansas State Board of Agriculture. 1887. *Fifth Biennial Report, for the Years 1885–86.*

Kansas State Board of Agriculture. 1889. *Sixth Biennial Report, for the Years 1887–88.*

Kansas State Board of Agriculture. 1895. *Ninth Biennial Report, for the Years 1893 and 1894.*

Kansas State Board of Agriculture. 1961. *Kansas Agriculture: Centennial Report* (44th Report).

Kopsa, Wilda. 1968. "Sod Houses—A Part of Prairie Homesteading." In *Cuba, Kansas, 1868–1968.* Sponsored by the Sunflower Kansas Extension Homemakers Unit. Pamphlet.

Laing, Francis S. 1910. "German-Russian Settlements in Ellis County, Kansas." *Collections of the Kansas State Historical Society (1909–1910)* 11:489–528.

Landes, Kenneth K. 1930. *The Geology of Mitchell and Osborne Counties, Kansas.* Kansas Geological Survey, Bull. 16.

Landes, Kenneth K. 1937. *Mineral Resources of Kansas Counties.* Kansas Geological Survey, Mineral Resources Circular 6.

Landes, Kenneth K., and Keroher, Raymond P. 1938. *Geology and Oil and Gas Resources of Rush County.* Kansas Geological Survey, Mineral Resources Circular 4.

Larkin, Arthur. 1965. "Manor House Life in Kansas." *Kansas City Star* May 17.

Latta, Bruce F. 1950. *Geology and Ground-water Resources of Barton and Stafford Counties, Kansas.* Kansas Geological Survey, Bull. 88.

Leonard, Alvin R., and Berry, Delmar W. 1961. *Geology and Ground-water Resources of Southern Ellis County and Parts of Trego and Rush Counties, Kansas.* Kansas Geological Survey, Bull. 149.

Lindquist, Emory. 1956. "The Scandinavian Element in Kansas." In *Kansas: The First Century,* ed. John D. Bright, pp. 307–328. New York: Lewis Historical Publishing Co.

Lindquist, Emory. 1963. "The Swedes in Kansas." *Heritage of Kansas* 7 (no. 1): 7–28.

Lindquist, Emory. 1963. "The Swedish Immigrant and Life in Kansas." *Kansas Historical Quarterly* 29:2–24.

Lingg, Warren J. 1961. *Early Days of Cawker City, Kansas.* Cawker City: Cawker City Ledger.

Logan, W. N. 1897. *The Upper Cretaceous of Kansas.* Introduction by Erasmus Haworth. University Geological Survey of Kansas 2:195–234.

Mack, Leslie. 1962. *Geology and Ground-water Resources of Ottawa County, Kansas.* Kansas Geological Survey, Bull. 154.

McLaughlin, T. J. 1949. *Geology and Ground-water Resources of Pawnee and Edwards Counties, Kansas.* Kansas Geological Survey, Bull. 80.

McNellis, J. M. 1973. *Geology and Ground-water Resources of Rush County, Kansas.* Kansas Geological Survey, Bull. 207.

Malin, James C. 1961. "Kansas: Some Reflections on Culture Inheritance and Originality." *Journal of the Central Mississippi Valley American Studies Association* 2:3–19.

Marsh, Susan. 1964. "Kansas Fenced in by Personalities of Stone." *New York Times* July 12.

[Martin, Geo. W.?]. 1900. "George A. Crawford." *Transactions of the Kansas State Historical Society (1897–1900)* 6: 237–248.

Mead, James R. 1906. "The Saline River Country in 1859." *Collections of the Kansas State Historical Society (1905–1906)* 9:8–19.

Mechem, Kirke. 1949. "Home on the Range." *Kansas Historical Quarterly* 17: 313–339.

Mechem, Kirke, ed., and Owen, Jennie Small, annalist. 1956. *The Annals of Kansas: 1886–1925,* vol. 2. Topeka: Kansas State Historical Society.

Mehl, Jane. 1970. *A Record of the Descendants of George Mehl and Mary Marquette Mehl, John Goodridge and Sarah Peavey Goodridge and John C. Peavey and Sarah W. Goodridge Peavey.* Cawker City: Cawker City Ledger.

Millbrook, Minnie Dubbs. 1955. *Ness: Western County, Kansas.* Detroit: Millbrook Printing Company.

Miller, Nyle H.; Langsdorf, Edgar; and Richmond, Robert W. 1961. *Kansas: A Pictorial History.* Topeka: Kansas Centennial Commission and State Historical Society.

Mooney, Mrs. Seymour. Letters to *Plattsburgh Republican* dated February 14, 1879; August 18, 1887; November 18, 1888; February 13, 1889; January 29, 1901. (From transcripts provided by Minnie Dubbs Millbrook.)

Morgenstern, William. 1923. "The Settlement of Bessarabia, Russia, by the Germans." *Collections of the Kansas State Historical Society (1919–1922)* 15:579–590. Translated from the German by J. C. Ruppenthal.

Moss, Rycroft G. 1932. *The Geology of Ness and Hodgeman Counties, Kansas.* Kansas Geological Survey, Bull. 19.

Mudge, B. F. 1878. "Geology of Kansas." *Kansas State Board of Agriculture First Biennial Report* 46–88.

Muilenburg, Grace. 1953. *The Kansas Scene.* Lawrence: Kansas Geological Survey.

Muilenburg, Grace. 1958. *The Land of the Post Rock.* 2nd ed. Mimeographed. Lawrence: Kansas Geological Survey.

Muilenburg, Grace. 1961. "Featuring the Kansas Landscape." Pamphlet 1 of *Stories of Resource-full Kansas.* Lawrence: Kansas Geological Survey.

"New Courthouse Buildings." 1958. *Kansas Government Journal* August.

New Scandinavia's Ninety-three Years (1868–1961): From Indian Days to Space Dreams. 1961. Compiled by Mrs. Homer Cardwell, Mrs. Joseph Johnson, and Mrs. Raymond Cooper.

Opdycke, Elizabeth. 1971. "Senior Citizen Dedicates Autobiography To Her Daughter; Records Hardships Endured by Early Pioneers in Kansas." *Russell Daily News* June 1.

Peters, Sarah. 1955. "Kansas has Unique Monument in the Stone Fenceposts of the Smoky Hills." *Kansas City Star* February 6.

Poell, Father John M. n.d. *The Cross in the Valley.* Pfeifer, Kansas: Holy Cross Church. Offset.

Prairiesta-100: Russell, Kansas. [1971]. Russell's centennial booklet.

Prentis, Noble L. 1899. *A History of Kansas.* Winfield, Kansas: E. P. Greer.

Raish, Marjorie Gamet. 1947. *Victoria: The Story of a Western Kansas Town.* Fort Hays Kansas State College Studies, General Series 12, Language and Literature Series 3.

Reeside, J. B. 1957. "Paleoecology of the Cretaceous Seas of the Western Interior of the United States." In *Treatise on Marine Ecology and Paleoecology*, vol. 2, *Paleoecology*, ed. H. S. Ladd. Geological Society of America, Mem. 67, part 2: 505–541.

Richards, W. M. 1960. "Fencing the Prairies." *Heritage of Kansas* 4 (no. 2): 7–18.

Richmond, Robert W. 1956. "Cowtowns and Cattle Trails." In *Kansas: The First Century*, ed. John D. Bright, 255–279. New York: Lewis Historical Publishing Co.

Risser, Hubert E. 1960. *Kansas Building Stone.* Kansas Geological Survey, Bull. 142:53–122.

Robbins, Roy M. 1942. *Our Landed Heritage: The Public Domain, 1776–1936.* Princeton: Princeton University Press. Reprint, Lincoln: University of Nebraska Press, A Bison Book, 1962.

Ross, Harry E. 1937. *What Price White Rock: A Chronicle of Northwestern Jewell County.* Burr Oak, Kansas: The Herald Press.

Rubey, W. W., and Bass, N. W. 1925. *The Geology of Russell County, Kansas.* Kansas Geological Survey, Bull. 10.

Ruede, Howard. 1937. *Sod-House Days: Letters from a Kansas Homesteader, 1877–78.* Edited by John Ise. New York: Columbia University Press.

Ruppenthal, Jacob C. 1915. "The German Element in Central Kansas." *Collections of the Kansas State Historical Society (1913–1914)* 13:513–534.

"Russians in Kansas." 1895. *Kansas City Journal* August 26.

St. Fidelis Church: "Cathedral of the Plains." n.d. Sponsored by Lions Club of Victoria, *et al.* Pamphlet.

Sanborn, Theo. A. 1973. "The Story of the Pawnee Indian Village in Republic County, Kansas." *Kansas Historical Quarterly* 39:1–11.

Sanders, John E., and Friedman, Gerald M. 1967. "Origin and Occurrence of Limestones." In *Developments in Sedimentology* vol. 9A: *Carbonate Rocks: Origin, Occurrence and Classification,* eds. George V. Chilingar, H. J. Bissell, and R. W. Fairbridge, 169–265. Amsterdam and New York: Elsevier Publishing Co.

Savage, I. O. 1901. *A History of Republic County, Kansas.* Beloit, Kansas: Jones & Chubbic.

Schmidt, C. D. 1906. "Reminiscences of Foreign Immigration Work for Kansas. *Collections of the Kansas State Historical Society (1905–1906)* 9:485–497.

Schoewe, Walter H. 1952. *Coal Resources of the Cretaceous System (Dakota Formation) in Central Kansas.* Kansas Geological Survey, Bull. 96:69–156.

Schoewe, Walter H. 1949. "The Geography of Kansas: Part 2, Physical Geography." *Kansas Academy of Sciences, Transactions* 52:261–333.

Skaggs, Jimmy M. 1973. *The Cattle-Trailing Industry: Between Supply and Demand, 1866–1890.* Lawrence: The University Press of Kansas.

Socolofsky, Homer E., and Self, Huber. 1972. *Historical Atlas of Kansas.* Norman: University of Oklahoma Press.

State of Kansas. 1868. *General Statutes.* Chapter 40, section 2, page 495.

State of Kansas. 1868. *General Statutes.* Chapter 105, section 12, page 1016.

Stoddard, Henry Luther. 1946. *Horace Greeley: Printer, Editor, Crusader.* New York: G. P. Putnam's Sons.

Swehla, Francis J. 1915. "Bohemians in Central Kansas." *Collections of the Kansas State Historical Society (1913–1914)* 13:469–501.

Toepfer, Amy Brungardt, and Dreiling, Agnes. 1966. *Conquering the Wind.* Garfield, New Jersey: Victor C. Leiker. (Printed at Tabloid Lithographers, Inc.)

Tolbert, Agnes. 1963. *Rock Houses of Minersville.* Chicago: Adams Press.

Tolsted, Laura Lu, and Swineford, Ada. 1957. *Kansas Rocks and Minerals.* 3d ed. Lawrence: Kansas Geological Survey.

T[ownsley], R[ussell] T. 1971. Editorial. *Russell Daily News* June 1.

Trewartha, Glenn T. 1941. "Climate and Settlement of the Subhumid Lands." In *Climate and Man: Yearbook of Agriculture.* U.S. Department of Agriculture: 167–176.

Van Deusen, Glyndon G. 1953. *Horace Greeley: Nineteenth-Century Crusader.* Philadelphia: University of Pennsylvania Press.

Waite, H. A. 1947. *Geology and Groundwater Resources of Ford County, Kansas.* Kansas Geological Survey, Bull. 43.

Walbridge, Caroline K. 1966. *Ranchorama and Louie C. Walbridge.* Russell: The Russell Record.

Waters, L. L. 1950. *Steel Trails to Santa Fe.* Lawrence: University of Kansas Press.

Webb, Walter Prescott. 1931. *The Great Plains.* Boston: Ginn and Company.

Webster, Mrs. Allen. n.d. Unpublished genealogy of the Faulhaber and Webster families. (Copy provided by Mrs. Martin DuVall, Salina.)

Wedel, Waldo R. 1959. *An Introduction to Kansas Archeology.* Smithsonian Institution Bureau of American Ethnology, Bull. 174. Washington, D.C.: U.S. Government Printing Office.

Whittemore, Margaret. 1954. *Historic Kansas: A Centenary Sketchbook.* Lawrence: University of Kansas Press.

Wilder, D. W. 1886. *The Annals of Kansas: New Edition, 1541–1885.* Topeka: T. Dwight Thacher, Kansas Publishing House.

Wing, Monta E. 1930. *The Geology of Cloud and Republic Counties, Kansas.* Kansas Geological Survey, Bull. 15.

Winsor, M., and Scarbrough, James A. 1878. "Jewell County." Reprint. *Collections of the Kansas State Historical Society (1926–1928)* 17:389–409, 1928.

Winther, Oscar O. 1971. "The English and Kansas, 1865–1890." In *The Frontier Challenge: Responses to the Trans-Mississippi West,* ed. John G. Clark, 235–273. Lawrence: The University Press of Kansas.

Zornow, William Frank. 1957. *Kansas: A History of the Jayhawk State.* Norman: University of Oklahoma Press.

NEWSPAPERS, PERIODICALS, AND MISCELLANEOUS MATERIALS EXAMINED

(Though only items in issues for dates italicized have been referred to in the text, something in most issues examined provided background or supporting information.)

NEWSPAPERS AND PERIODICALS

Atchison Champion. July 16, 1881.

Bazine Advocate. February 19 and May 28, 1937.

Belleville Telescope. New Year's edition, December 29, 1905. Scattered issues, 1930s.

Beloit Courier. May 1, 1879.

Beloit Daily Call. July 26, 1934. Scattered issues, late 1950s through 1960s.

Beloit Gazette. December 27, 1879. Scattered issues, 1880 and 1881. March 2 and 9, 1961.

Beloit Weekly Record. November 23, 1877.

Bunker Hill Advertizer. Scattered issues, 1870s; 1880 and 1881. November 23 and September 7, 1933; October 10, 1935.

Bunker Hill Banner. March 17, 1882.

Burlington Weekly Hawkeye. July 1, 1882 [or 1883].

Burr Oak Herald. Several issues, 1936.

Cawker City Ledger. July 18, 1935. Scattered issues.

Clyde Republican. Scattered issues.

Concordia Blade. July 9, 1935.

Concordia Blade-Empire. Scattered issues.

The Cooperative Consumer. Article on Post Rock Museum by Marguerite Kingman, April 15, 1966.

Courtland Journal. September 15, 1935; September 8 and 15, 1960.

Delphos Republican. July 12, 1962. Scattered issues.

The Denver Post. Article on fence posts by Charlotte Norlin, April 16, 1967.

The Diamond (Jewell City ?). July 15, 1876.

Dodge City Daily Globe. January 22 and 23, 1935. Article on Margaret Haun Raser by Ethel Watkins, February 20, 1954. Scattered issues.

Downs News. Scattered issues.

Ellis County News. Several issues, August and September, 1926; October 11 and 18, 1951.

Ellis County Star. June 21, 1880.

Ellis Review. August 6, 1970.

Ellsworth Messenger. Scattered issues in the 1950s (especially historical articles by George Jelinek).

Ellsworth Reporter. March 11 and 18, 1880. Scattered issues.

Farmland News. Article on "plains cathedral" by Ed Duckworth, May 31, 1972.

Glasco Sun. Scattered issues.

Glen Elder Sentinel. Several issues, April, May, June, 1933.

Great Bend Tribune. December 11, 1934; January 26 and 27, 1935. Scattered issues.

Hays Daily News. Scattered issues between May 23, 1933, and October 8, 1942. Series on Victoria appearing periodically between September 21, 1950, and November 9, 1958. January 6, 1961; September and October, 1964; February 18, 1968; July 26, 1970; April 19, 1971.

Hays Sentinel. Several issues, 1877; February 8, 1878.

High Plains Journal. Scattered issues.

Hoisington Dispatch. August 13 and September 10, 1936. Scattered issues.

Holyrood Gazette. Scattered issues.

Hutchinson News. January 6, 1946; January 6, 1957; February 2, 1958; September 28, 1962; May 17 and December 6, 1964; June 6, 1971; January 20, 1973.

Jetmore Republican. April 25, 1930; May 16 and 23, 1957. Scattered issues.

Jewell County Record (Mankato). Scattered issues.

Jewell County Republican (Jewell). April 28, 1933; January 29, 1953.

The Kansan (Concordia). May 7, 1935.

Kansas! (Kansas Department of Economic Development). Most issues.

Kansas City Journal. August 26, 1895.

Kansas City Star. September 23, 1903; January 12, 1907; *August 27* and *October 22, 1911*; August 11, 1928; August 15, 1943; May 5, 1963. Article on "Cathedral of the Plains" by Helen Huyck, January 4, 1972.

Kansas Farmer. August 4, 1973. Scattered issues.

Kansas Fish and Game (Kansas Forestry, Fish and Game Commission). Many issues.

Kansas Optimist (Jamestown). Scattered issues.

Kansas Stockman. May 1973.

La Crosse Chieftan. September 3, 1936.

La Crosse Republican. April 1, 1937.

Lincoln Beacon. March 3, 1887.

Lincoln County News. October 6, 1932; May 11, 1933.

Lincoln Republican. March 26, 1917.

Lincoln Sentinel. Issues for *November, 1933*; December 14, 1934; August 8, 1946.

Lincoln Sentinel-Republican. May 3, 1956. Scattered issues since.

Lucas Independent. Scattered issues.

Lucas-Sylvan News. Scattered issues.

Mankato Advocate. July 23, 1940.

Minneapolis Messenger. Issues for August 1966.

Natoma-Luray Independent. Scattered issues.

Ness County News. June 8, 1935; May 30, 1946; June 2, 1966. Scattered issues.

Osborne County Farmer. January 1, 1880; December 25, 1913; December 3, 1936; July 28, 1966; July 8 and 15, 1971. Scattered issues.

Paradise Farmer. February 5 and June 25, 1934.

Public Record (Cawker City). *Supplement 1885.*

Rush County News. May 17, 1951. Series of articles on Rush County Historical Society's project, Post Rock Museum, between July 1962 and May 1964.

Russell County News. February through March, 1932; December 26, 1935.

Russell [County] Record. Issues for 1875 and 1876; *March 27, 1879; February 28, 1884; February 10, 1900; September 27, 1902*; March 14, 1906; December 21. 1922; January and April through June, 1933; *May 8, 1941*; May 24, 1951. Scattered issues, 1960s.

Russell Daily News. June 3, 1961; June 1, 1971. Scattered issues.

Salina Journal. Scattered issues, late 1950s through 1960s. *July 31, 1971.*

Scandia Journal. Scattered issues.

Simpson News. April 4 and November 28, 1935.

Sylvan Grove News. Scattered issues.

Star (Sunday magazine of the *Kansas City Star*). Article on "Cathedral of the Plains" by David Dary, December 17, 1972.

Topeka Commonwealth. February 25, 1872.

Topeka Daily Capital. June 10, 1890; June 18, 1911; July 26, 1936; August 31, 1941. Article on "Cathedral of the Plains" by James Robinson, April 16, 1954. *September 30, 1956.* Article on Cawker City by Peggy Greene, May 31, 1959. May 5, 1963.

The Visitor (National Catholic weekly). Article on Catholic churches in Ellis County by Margaret Hale Malsam, March 5, 1972.

Waldo Advocate. February 5, 1934; January 14, 1935.

Walnut City [Rush Center] *Herald. March 17, 1886.*

Washington County News. Scattered issues.

Washington County Register. February 24 and March 17, 1933. January 11 to July 17, 1935.

The Wichita Eagle and Beacon. November 11, 1963; August 30, 1964. Scattered issues, especially articles by Forrest Hintz on central Kansas historic and scenic features.

Wilson Echo. Scattered issues, 1880.

Wilson Record. January 25, 1933.

Wilson World. July 22, 1929; January 25, 1933; February 13, 1935; July 1948. Scattered issues to date.

MISCELLANEOUS

Brochures, leaflets, and maps distributed by U.S. Bureau of Reclamation; U.S. Corps of Engineers; Kansas Department of Economic Development; Kansas Park and Resources Authority; Kansas Highway Commission; and area Chambers of Commerce, civic groups, historical societies, and other groups.

County clippings (in scrapbooks), Kansas State Historical Society.

J. C. Ruppenthal's columns on local history, appearing from the 1920s to the 1950s in area newspapers, especially in Russell, Ellsworth, and Lincoln counties.

SPECIAL SOURCES

The following contributors, by interview and correspondence, provided basic information over an extended time:

* C. C. Abercrombie, Barnard
Bird Abram, Beloit
Matilda Adams, Lyons
William Appel, La Crosse
William Baier, Victoria
Don D. Ballou, Kansas City, Kansas
Olive (Mrs. Jesse) Barker, Salina
Wayne Barnett, Glen Elder
Mrs. Ida Brown, Lincoln
The Rev. Blaine E. Burkey, Hays
Margaret Evans Caldwell, Hanston
Myron and Bernice Chapman, Beloit
Ralph Coffeen, Russell
Helen Craig Dingler, Enterprise
John A. Dinkel, Victoria
* George Doctor, Belleville
Ollie (Mrs. George) Doctor, Belleville
Philip J. Doyle, Beloit
Floyd Duncan, Manhattan
Donna (Mrs. Martin) DuVall, Salina
Harold Dwyer, Hastings, Nebraska
Roy Ehly, La Crosse
The Rev. Raphael Engel, Hays
Inez Ernzen, Beloit
Kate Lee (Mrs. R. N.) Ewing, Russell
Peter F. Felten, Jr., Hays
Mrs. Hazel Flaherty, Lincoln
* Thomas G. Frusher, Tulsa, Oklahoma
Gertrude (Mrs. Loren) Furney, Russell
Mrs. Ralph Goodheart, Lucas
Harry Grass, La Crosse
Paul V. Grittman, Simpson
C. V. Haggman, M. D., Scandia
The Rev. Jordan S. Hammel, Victoria
Mrs. Margaret Hanni, Beloit
Ernest Hanzlicek, Wilson
Donald E. Hattin, Bloomington, Indiana
Ephriam Hedstrom, Courtland
J. M. (Jake) Herrman, Liebenthal
Agnes (Mrs. John) Hill, Wilson
* John Hill, Wilson
Mrs. C. Hobbs, Lincoln
R. D. (Rex) Hodler, Beloit
Helen M. Holland, Russell
Alan Houghton, Beloit
C. R. Hubbard, Beloit
Gene Humburg, Ness City

Harold Humburg, Ness City
George Jelinek, Ellsworth
Burnice (Mrs. George) Jensen, Jamestown
Mr. and Mrs. Arthur Jepsen, Lincoln
Hans Jorgensen, Lincoln
George J. Klaus, Hays
Bessie (Mrs. George) Lang, Cuba
Mrs. Blanche Lange, Bunker Hill
Herman Linenberger, Victoria
Mrs. Robert J. Mahoney, Bunker Hill
M. K. Mathews, Quinter
Emil Mauth, Bazine
Clarence B. Mehl, Jr., Beloit
David Mehl, Beloit
Jane Mehl, Beloit
Robert E. Meneley, Russell
Minnie Dubbs Millbrook, Topeka
Mr. and Mrs. Golden Morris, Lincoln
Mabel (Mrs. Bert) Oakley, Asherville
* Clarence Peck, Bunker Hill
The Rev. John M. Poell, Pfeifer
Joe Prickett, Beloit
* Margaret Haun Raser, Jetmore
Kirk C. Raynesford, Big Bear Lake, California
Harvey Roush, Lincoln and Ellsworth
Herbert W. Sandell, Manhattan
Earle Sawyer, Lucas
Johnnie Schmitt, Gorham
* Charles Scholer, Manhattan
Margaret (Mrs. Melvin) Schulz, Hunter
Parker V. Seirer, WaKeeney
O. H. Senter, Beloit
Mrs. Leland Shaffer, Bunker Hill
Ray Shaffer, Bunker Hill
* Edward W. Shulda, Belleville
Winifred A. Smies, Courtland
Charles Speck, Sterling
Edythe Raynesford (Mrs. Charles) Speck, Sterling
Charles A. Studt, Glasco
George Thielen, Dorrance
Russell T. Townsley, Russell
Mrs. Adolph Vopat, Wilson
Myrl V. Walker, Hays
* Mrs. Allen Webster, Lincoln
Fred Wessling, Beloit
Charles White, Paradise
Clarence Youse, Dodge City
Mrs. Emil Zahradnik, Wilson
Mary Zajic and Derald Zajic, Holyrood

These contributors provided supporting information, verified copy, or directed us to fruitful sources:

Mrs. C. C. Abercrombie, Barnard
Mrs. Helen Abercrombie, Barnard
Fred Abram, Clyde
Bertha Ackerman, Lincoln
Paul Adams, Oklahoma City, Oklahoma
Winstan Anderson, M. D., Lawrence
Mrs. Essie Ballou, Delphos
R. L. Barnum, Glasco
Francis J. Baxa, Belleville
Georgiana (Mrs. Frank) Bell, Russell
Jack Bennett, Beloit
Mr. and Mrs. Albert Benson, Lincoln
The Rev. W. L. Birzer, Liebenthal
Mr. and Mrs. Ben Blecha, Munden
Leo Boutz, Concordia
* Mamie (Mrs. Frank W.) Boyd, Mankato
Mary (Mrs. Frank, Jr.) Boyd, Mankato
Robert R. Bruegger, Hays
Ray Buck, Lincoln
Mr. and Mrs. Clyde Bunker, Hays
H. C. Cain, Delphos
Wilson Carlgren, Concordia
Marvin P. Carlson, Lincoln, Nebraska
John H. Carson, Miltonvale
Grace (Mrs. Byron) Chegwidden, Bunker Hill
J. R. Chelikowsky, Manhattan
Ray L. Claycomb, Pipestone, Minnesota
Velma C. Cooper, Lyons
Mrs. Melvern Cornett, Belleville
Keith Cossman, Jetmore
Herman Crawford, Lincoln
Agnes and Nora Dahl, Topeka
Loren Dees, Ellsworth
* Jess Denious, Jr., Dodge City
Ben Doctor, Lawrence
Lawrence Dooley, Scandia
Frank Dowlin, Barnard
W. W. Duitsman, Hays
Mrs. M. S. Duvall, Hoisington
Ernest J. Elniff, Randall
Frances Walter (Mrs. Ed) Elyea, Belleville
Mr. and Mrs. John Errebo, Lincoln
Allan D. Evans, Russell
George F. Ferguson, Beverly
Mrs. Vernon Felder, Nekoma
Melvin Flegler, Russell
Robert Frusher, Jetmore
Mrs. Edna Garrett, Dodge City
Clara Gates, Mankato
Margaret Gentry, Beloit
Barbara (Mrs. Harry) Grass, La Crosse

Charles L. Hall, Manhattan
Helen L. Hall, Hutchinson
Jeanne Hanni, Beloit
Mrs. Lola Harper, Dodge City
Winston H. Hedges, McCook, Nebraska
Mrs. Paul W. Heiman, Beloit
Andrew Herrman, Liebenthal
Mrs. Charles Hill, Wilson
Bert Hitchcock, Russell
Madge Seirer (Mrs. Clyde) Hooper, Sylvan Grove
Loren Howe, Belleville
Clara Williams Humburg, La Crosse
Neil Humburg, Topeka and Newton
David Jeffries, La Crosse
Mr. and Mrs. Martin R. Jensen, Lincoln
Mrs. J. Rudolph Johnson, Courtland
Louise (Mrs. W. Carl) Johnson, Salina
Ross Kelly, Glasco
Mr. and Mrs. Earl Keeler, Barnard
Mrs. La Vern Kopsa, Cuba
Ivan Krug, La Crosse
Ed J. Kuhn, Victoria
Edward L. Kunze, Scandia
Warren Lingg, Cawker City
Earl and Elaine Loganbill, Beloit
Mr. and Mrs. Howard Mahin, Randall
William Mathews, Vancouver, British Columbia
Jerry Maxfield, Midland, Texas
Mrs. Elizabeth McCoy, Beloit
Mrs. Larry Meeks, Olmitz
Alice Meier, Lincoln
Mrs. Mildred Middlekauff, Sylvan Grove
Mrs. Albina Miller, Victoria
Virginia Montgomery, Barnard
Mrs. Lois Muchow, Brookville
Mr. and Mrs. Wayne Naegele, Lucas
Anita F. and Wilbur Oetting, Slyvan Grove
Mr. and Mrs. Verdon Peckham, Hunter
Robert (Bob) Perske, Omaha, Nebraska
Ruth (Mrs. Robert) Pickrell, Salina
Mr. and Mrs. Ed Popelka, Cuba
Mr. and Mrs. George Putt, Mankato
Margaret (Mrs. Victor) Randolph, Osborne
Ray Rasmussen, Lincoln
Elmer and Evelyn Rebenstorf, Sylvan Grove
Lafe Rees, Lincoln
Robert W. Richmond, Topeka
Mrs. Joe Riedl, Hoisington
D. E. Robinson, Topeka
Trohman Robinson, Dodge City
Jane Robison, Dodge City

Charles B. Rogers, Ellsworth
Mrs. Walter Rothe, Ness City
Warren Rude, Hoisington
Frank Salmans, Jetmore
Mr. and Mrs. Celestine Sander, Victoria
B. J. Schneider, Beloit
Mr. and Mrs. Glenn Schniepp, Bazine
Leo Schugart, Hoisington
Gilbert Schwartz, Gorham
Denis Shumate, Beloit
George Sis, Belleville
The William Smies family, Courtland
Thayer Smith, Salina
E. J. Snyder, Holyrood
Mr. and Mrs. Joe Sterba, Cuba
Mr. and Mrs. Frank Stouffer, Beloit
Waldimore Strecker, Bazine
A. L. Street, Beloit
Lois E. Strnad, Munden
Earl Thomas, Narka
Mrs. Frances K. Thull, Cawker City
Warren Tindall, Hoisington

Mark Torr, Salina
Mrs. Alma Vaughn, La Crosse
O. H. Wagner, Mankato
Caroline Walbridge, Topeka
Albert Walker, Lincoln
Frank B. Walker, Lincoln
Leo Wallace, Barnard
Tressa (Mrs. Wilbur) Wallace, Barnard
Larry Walter, Belleville
Mrs. Ellen Warren, Troy
Everett Waudby, Gorham
Mr. and Mrs. Frank Webster, Hunter
George Welling, Paradise
Robert Wenthe, Sylvan Grove
The Rev. Loren Werth, Concordia
Ralph West family, Narka
Carl Westin, Mankato
Don Wigington, Quinter
Mrs. Frank Williams, Glasco
Francis Wilson, Ellsworth
Thomas Witty, Topeka
Dwight A. Yordy, Brookville

Deceased

Index

References to illustrations are printed in italic type

191